Industrial Approaches in Vibration-Based Condition Monitoring

Industrial Approaches in Vibration-Based Condition Monitoring

Jyoti K. Sinha

CRC Press
Taylor & Francis Group
Boca Raton London New York

CRC Press is an imprint of the
Taylor & Francis Group, an **informa** business

CRC Press
Taylor & Francis Group
6000 Broken Sound Parkway NW, Suite 300
Boca Raton, FL 33487-2742

© 2020 by Taylor & Francis Group, LLC
CRC Press is an imprint of Taylor & Francis Group, an Informa business

No claim to original U.S. Government works

Printed on acid-free paper

International Standard Book Number-13: 978-1-138-55033-9 (Hardback)

This book contains information obtained from authentic and highly regarded sources. Reasonable efforts have been made to publish reliable data and information, but the author and publisher cannot assume responsibility for the validity of all materials or the consequences of their use. The authors and publishers have attempted to trace the copyright holders of all material reproduced in this publication and apologize to copyright holders if permission to publish in this form has not been obtained. If any copyright material has not been acknowledged please write and let us know so we may rectify in any future reprint.

Except as permitted under U.S. Copyright Law, no part of this book may be reprinted, reproduced, transmitted, or utilized in any form by any electronic, mechanical, or other means, now known or hereafter invented, including photocopying, microfilming, and recording, or in any information storage or retrieval system, without written permission from the publishers.

For permission to photocopy or use material electronically from this work, please access www.copyright. com (http://www.copyright.com/) or contact the Copyright Clearance Center, Inc. (CCC), 222 Rosewood Drive, Danvers, MA 01923, 978-750-8400. CCC is a not-for-profit organization that provides licenses and registration for a variety of users. For organizations that have been granted a photocopy license by the CCC, a separate system of payment has been arranged.

Trademark Notice: Product or corporate names may be trademarks or registered trademarks, and are used only for identification and explanation without intent to infringe.

Library of Congress Cataloging-in-Publication Data

Names: Sinha, Jyoti Kumar, author.
Title: Industrial approaches in vibration-based condition monitoring / Jyoti Sinha.
Description: Boca Raton : CRC Press, 2020. | Includes bibliographical
references and index.
Identifiers: LCCN 2019038663 | ISBN 9781138550339 (hardback) | ISBN
9781315147222 (ebook)
Subjects: LCSH: Vibration--Testing. | Vibration--Measurement. |
Machinery--Monitoring. | Nondestructive testing.
Classification: LCC TA355 .S524 2020 | DDC 620.3/7--dc23
LC record available at https://lccn.loc.gov/2019038663

Visit the Taylor & Francis Web site at
http://www.taylorandfrancis.com

and the CRC Press Web site at
http://www.crcpress.com

Dedicated to my father

Mr. (Late) Jagdish Prasad Sinha
(7 February 1936 to 27 January 2018)

Contents

Preface xiii
Author xv

Chapter 1 Introduction 1

    1.1   Condition Monitoring 1
    1.2   Condition Monitoring Techniques 1
    1.3   Condition-Based Maintenance 4
           1.3.1   Lead-Time-to-Maintenance (LTM) 5
    1.4   Summary 9

Chapter 2 Simple Vibration Theoretical Concept 11

    2.1   Equation of Motion 11
    2.2   Damped System 13
           2.2.1   Equation of Motion for Free Vibration 14
           2.2.2   Critically Damped System 15
           2.2.3   Over Damped System 15
           2.2.4   Under Damped System 16
    2.3   Forced Vibration 18
           2.3.1   Example 2.1: An SDOF System 22
    2.4   Concept of Modeshapes 23
    2.5   Machine Vibration 25
           2.5.1   Rotor Dynamics 25
           2.5.2   Unbalance Responses 27
           2.5.3   Machine Faults 32
    2.6   Summary 33
    References 33

Chapter 3 Vibration-Based Condition Monitoring and Fault Diagnosis:
Step-by-Step Approach 35

    3.1   Introduction 35
    3.2   Different Stages of Vibration Measurements and
Monitoring 35
           3.2.1   Bath Tub Concept 35
           3.2.2   Stage 1—Machine Installation and Commissioning 35
           3.2.3   Stage 2—Machine Operation 38
           3.2.4   Stage 3—Aged Machines 42
    3.3   Summary 43
    References 43

vii

viii   Contents

Chapter 4 Vibration Instruments and Measurement Steps 45

       4.1   Introduction 45
       4.2   Sensors and Their Mounting Approach 45
               4.2.1   Displacement Sensor..45
               4.2.2   Velocity Sensor...47
               4.2.3   Acceleration Sensor..48
               4.2.4   Tacho Sensor 53
       4.3   Vibration Measurement 55
               4.3.1   A Typical Measurement Setup 55
               4.3.2   Steps Involved in the Data Collection.........................57
               4.3.3   Instrument Calibration and Specifications...................59
               4.3.4   Concept of Sampling Frequency 61
               4.3.5   Aliasing Affect and Anti-aliasing Filter 61
                        4.3.5.1   Observations 65
               4.3.6   Concept of Nyquist Frequency, f_q and the Useful
                        Upper Frequency Limit, f_u.......................................66
               4.3.7   Analog-to-Digital Conversion (ADC)........................67
       4.4   Conversion of the Measured Data into the Mechanical Unit.....70
       4.5   Summary 71
       References 71

Chapter 5 Signal Processing 73

       5.1   Time Signal 73
               5.1.1   Filters..73
               5.1.2   Amplitude of Vibration 74
               5.1.3   Integration of Time Domain Signal 76
               5.1.4   Statistical Parameters...78
               5.1.5   Comparison between CF and Kurtosis 79
       5.2   Fourier Transformation (FT)..82
               5.2.1   Example 5.1: A Sine Wave Signal 83
               5.2.2   Steps Involved for the Computation of FT...............84
               5.2.3   Importance of Frequency Resolution in
                        Spectrum Analysis 87
               5.2.4   Leakage 88
               5.2.5   Window Functions 89
       5.3   Computation of Power Spectral Density (PSD) 92
               5.3.1   Averaging Process...92
               5.3.2   Concept of Overlap in the Averaging Process............94
               5.3.3   Example 5.2: An Experimental Rig 96
               5.3.4   Example 5.3: An Industrial Blower............................96
       5.4   Conversion of Acceleration Spectrum to Displacement
              Spectrum and Vice Versa 99
       5.5   Short Time Fourier Transformation (STFT) 100
               5.5.1   Example 5.4: An Experimental Rig 101
               5.5.2   Example 5.5: An Industrial Centrifugal Pump 101

Contents   ix

5.6		Correlation between Two Signals	103
	5.6.1	Cross Power Spectrum	103
	5.6.2	Transfer Function (Frequency Response Function)	104
	5.6.3	Ordinary Coherence	104
	5.6.4	Example 5.6: Two Simulated Signals with Noise	105
	5.6.5	Example 5.7: Laboratory Experiments	107
5.7		Concept of Envelope Analysis	109
5.8		Summary	111
References			111

Chapter 6   Vibration Data Presentation Formats 113

6.1		Introduction	113
6.2		Normal Operation Condition	113
	6.2.1	Overall Vibration Amplitude	113
	6.2.2	Vibration Spectrum	113
	6.2.3	The Amplitude—Phase versus Time Plot	115
	6.2.4	The Polar Plot	115
	6.2.5	The Orbit Plot	116
6.3		Transient Operation Conditions	117
	6.3.1	The 3D Waterfall Plot of Spectra	117
	6.3.2	The Shaft Centerline Plot	118
	6.3.3	The Orbit Plot	118
	6.3.4	The Bode Plot	119
6.4		Summary	120
References			120

Chapter 7   Vibration Monitoring, Trending Analysis and Fault Detection 121

7.1		Introduction	121
7.2		Types of Faults	125
7.3		Rotor Faults Detection	125
	7.3.1	Mass Unbalance	125
	7.3.2	Shaft Bow or Bend	126
	7.3.3	Misalignment	127
	7.3.4	Shaft Crack	128
	7.3.5	Shaft Rub	128
7.4		Other Machine Fault Detection	129
	7.4.1	Mechanical Looseness	129
	7.4.2	Blade Passing Frequency (BPF)	129
	7.4.3	Blade Vibration and Blade Health Monitoring (BHM)	129
	7.4.4	Electric Motor Defects	129
7.5		Gearbox Fault Detection	130

7.6	Anti-friction Bearing Fault Detection	137
	7.6.1 Crest Factor (CF)	141
	7.6.2 Kurtosis (Ku)	141
	7.6.3 Envelope Analysis	142
7.7	Experimental Examples	143
	7.7.1 Example 7.1—Roller Bearing Defect	143
	7.7.2 Example 7.2—Rotor Faults	144
7.8	Industrial Examples	147
	7.8.1 Example 7.3—Fan with Unbalance Problem	147
	7.8.2 Example 7.4—Gearbox Fault	148
7.9	Machines Having Fluid Bearings	151
7.10	Field Rotor Balancing	153
	7.10.1 Single Plane Balancing—Graphical Approach	153
	7.10.2 Single Plane Balancing—Mathematical Approach	155
7.11	Summary	157
	References	157

Chapter 8 Experimental Modal Analysis 159

8.1	Experimental Procedure	159
	8.1.1 Impulsive Load Using the Instrumented Hammer	159
8.2	Modal Analysis	162
8.3	Experimental Examples	170
	8.3.1 Example 8.1—A Clamped-Clamped Beam	170
	8.3.2 Example 8.2—Experimental Rotating Rig-1	178
	8.3.3 Example 8.3—Experimental Rotating Rig-2	181
8.4	Industrial Examples	183
	8.4.1 Example 8.4—Horizontal Centrifugal Pump	183
	8.4.2 Example 8.5—Vertical Centrifugal Pump	186
	8.4.3 Example 8.6—Wind Turbine	188
8.5	Summary	191
	References	191

Chapter 9 Operational Deflection Shape (ODS) 193

9.1	Simple Theoretical Concept	193
9.2	Industrial Examples	198
	9.2.1 Example 9.1—Steam Turbo-Generator (TG) Set	198
	9.2.2 Example 9.2—Gearbox Failure	202
	9.2.3 Example 9.3—Blower with Frequent Bearing Failure	204
9.3	Summary	207
	References	207

Contents   xi

Chapter 10 Shaft Torsional Vibration Measurement 209

    10.1   Measurement Approach... 209
    10.2   Extraction of Torsional Vibration Signal.............................. 210
           10.2.1   Time Domain Zero-Crossing Approach 210
           10.2.2   Demodulation Approach 212
    10.3   Experimental Examples 213
           10.3.1   Example 10.1—Blade Vibration............................ 213
           10.3.2   Example 10.2—A Diesel Engine 216
    10.4   Summary 218
    References 218

Chapter 11 Selection of Transducers and Data Analyzer for a Machine............ 219

    11.1   Introduction 219
    11.2   Calculation of Machine Faults Frequencies 219
    11.3   Selection of Accelerometer... 221
    11.4   Analysis Parameters 221
           11.4.1   Time Domain Analyses.. 222
           11.4.2   Frequency Domain Analyses 222
           11.4.3   Time-Frequency Analyses..................................... 222
    11.5   Features Required in the Data Analyzer 222
           11.5.1   Specifications 223
           11.5.2   Data Analysis Capabilities 223
           11.5.3   Data Trending and Storage..................................... 224
    11.6   Summary 224

Chapter 12 Future Trend in VCM... 225

    12.1   Introduction 225
           12.1.1   Future IIoT-Based CVCM Approach 227
    12.2   Approach 1: Suitable for Existing Old Plants...................... 227
    12.3   Approach 2: Suitable for New Plants................................... 229
    12.4   Summary 230
    References 231

Index... 233

Preface

Technology is evolving rapidly. The Fourth Industrial Revolution is shaped by the growing ubiquity of Industrial Internet of Things (IIoT) and making use of tools, such as of AI and data analysis, to solve problems in a large scale. Our dependency on technology is also growing, but finding experts and those who understand the physics of the subject matter is important to get meaningful outputs from the technology.

Vibration-Based Condition Monitoring (VCM) is built on a set of techniques to interpret the data correctly, irrespective of how technologically advanced the VCM instruments and systems are in any industry. This book explores the VCM methods, with clear explanations, to develop the reader's understanding of the subject.

I'd like to acknowledge all of the co-authors of my several research publications. Some of their findings, results and figures are used in this book as examples.

Finally I would like to thank my late father Mr Jagdish Sinha, my mother Mrs Chinta Sinha, my wife Sarita, my son Aarambh and all of my family, friends, students and colleagues who have supported me to complete this book.

Jyoti K. Sinha

Author

Prof. Jyoti K. Sinha, BSc (Mech), MTech (Aero), PhD, CEng, FIMechE, is Programme Director of Reliability Engineering and Asset Management (REAM) MSc course, head of the dynamics laboratory, and head of structures, health and maintenance (SHM) research group, School of Engineering, The University of Manchester, Manchester, UK.

Prof. Sinha is an internationally well-known expert in vibration-based condition monitoring and maintenance of machines and structures. He is involved in and solved a number of industrial vibration problems of machines, piping and structures by *in-situ* vibration measurements and analysis in many plants over the last 30 years. Sinha is the author of more than 225 publications (journals, conferences, books, edited book/conference proceedings and technical reports) and gave a number of keynote/invited lectures. He is author of a book "Vibration Analysis, Instruments and Signal Processing" and co-author of two other books.

Prof. Sinha has started a series of international conferences of maintenance engineering (IncoME from August 2016). Prof. Sinha is also the associate editor of three international journals—*Structural Health Monitoring: An International Journal*; *Mechanism and Machine Theory*; and *Journal of Vibration Engineering and Technologies*. He is also an editorial board member of the journals *Structural Monitoring and Maintenance* and *Machines*, as well as a technical committee member of IFTOMM Rotordynamics and a member of several international conferences.
https://www.research.manchester.ac.uk/portal/Jyoti.Sinha.html.

1 Introduction

1.1 CONDITION MONITORING

It is well known that machines keep on developing faults (or defects) during their normal operation that often lead to failures if not monitored regularly. The failures are generally random and may occur in different components of any machines. Figure 1.1 shows random failures of a motor, bearing, rotor and other components in a machine. The probability of failure distribution for a machine can generally be represented by Gaussian distribution as shown in Figure 1.2. The mean-time-to-failure (MTTF) may be different for many identical machines. This simply means that there are many failures before the MTTF. It is also difficult to predict which component is going to fail next based on statistical data.

The impact of failures of any machine or any unit within an industry depends upon its functional requirements. The failure of any critical machines/equipment may impact plant safety, production loss and more maintenance overhead. The random up-and-down time of any critical machines or any plant (Figure 1.3) may not be acceptable. Hence a more reliable approach is needed that can monitor the health condition during the plant operation and predict the defect (or fault), if any, before failure. This process is known as "Condition Monitoring (CM)."

1.2 CONDITION MONITORING TECHNIQUES

There are several non-destructive test (NDT) techniques are available and currently used in industries to meet the need of the CM. A few of the commonly used techniques are listed here.

1. **Monitoring of machine and process parameters**: It is often found to be the best indicator of machine health condition and its performance e.g., car dashboard providing information of engine revolution per minute, coolant level and temperature, parking break, engine condition, etc.
2. **Temperature monitoring**: The temperature monitoring is also useful in many cases to identify the defect such as mechanical rubbing in machines, defects in anti-friction bearings, leak in pipe, etc.

   A temperature sensor can measure the temperature at a point of machine and structure. Hence more number of sensors is required to

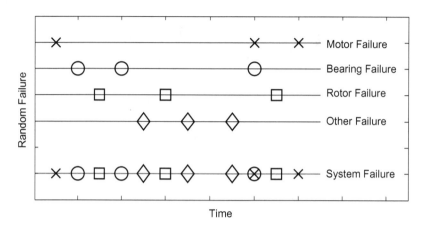

FIGURE 1.1 Random failures.

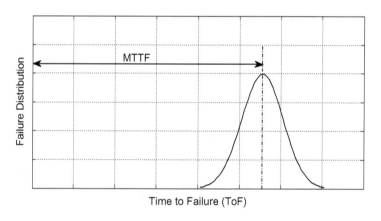

FIGURE 1.2 Probability distribution of failures.

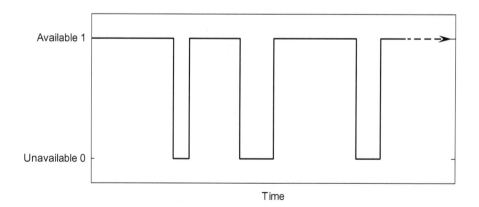

FIGURE 1.3 Up time and down time.

Introduction

map the temperature distribution of a machine or structural surface. **Thermography** overcomes this limitation. This works on the infrared energy emission concept to measure the object surface temperature. The thermography camera is now popular and commonly used to monitor the machines and structure surface temperature remotely. It can identify the "hot spot" quickly and useful for several applications. The simplest example is identifying the hot spot in the main electrical circuit board of any building (shopping mall, hospital, plant, etc.) due to looseness in electric circuits.

3. **Lubricant monitoring**: The lubricant is used in the several machines. The lubricant condition keeps on deteriorating during its use which may impact the machine performance; therefore its regular mentoring is required so that it can be replaced in time. The following two monitoring processes are useful.

   Lubricant properties – This is mainly related the monitoring of the concentration of additives and other properties of the lubricant.

   Debris analysis – It is measure of number of metal particles and their sizes in the lubricant. The wear particles from the rubbing surface in the lubricant can be used to identify the defects in bearings, gearboxes, etc.

4. **Leak detection monitoring**: The leak detection in the pipe conveying fluid is essential for several applications. Pressure sensor, acoustics wave reflection, etc. can be to detection the leak. The ultrasound technique may be useful but it is a time-consuming process for long pipeline. The tracer gas technique can also be used at certain time interval.

5. **Noise monitoring**: Microphones or hydrophones are used for this purpose depending upon the surrounding medium. This is found to be useful for monitoring machines, damage in pipelines, etc.

6. **Acoustic emission monitoring**: It is observed that the occurrence of crack, propagation of damage in any structures, bearing defects, etc. generally release energy at very high frequency range (more than 70–80 kHz). Such signals can be picked up by the acoustic emission sensor; however the detection process is heavily dependent on the resonance frequency of the sensor. Therefore an appropriate selection of the sensor is essential for any particular application.

7. **Vibration monitoring**: Vibration monitoring is widely used in the CM of machines and structures in industries. The vibration can detects the change in any objects (machines or structures) that can be explained through the mechanical vibration theory; hence, the vibration measurements, tests and analysis are already popular tools for several applications such as design qualification, design optimization, condition monitoring of structures and machines, vibration insolation and control, etc.

Vibration measurements and data analysis related to machines are mainly discussed in this book, but the same instruments and concepts can be applied to any structural components.

1.3 CONDITION-BASED MAINTENANCE

A range of maintenance strategies is available to the maintenance manager. They are:

1. **Breakdown Maintenance (BM) or Run to failure**: Here the maintenance activity is carried out after the failure of items. It is good to select those items of the plant that must have minimal impact on plant safety and production, e.g., electric bulbs.
2. **Planned Preventive Maintenance (PPM)**: This approach is useful for those items that have predictive life; hence these items can be replaced at a fixed time interval related to their life. But this approach may have risks, such as the item may be replaced too soon, item may fails before replacement, etc.
3. **Condition-based Maintenance (CBM)**: This is a much better approach to avoid failure by detecting defect at the early stage through the CM. The predication of the defect or degradation or deterioration by the CM can triggers the maintenance activity to rectify the defect; hence the approach is known as the CBM (or Predictive Maintenance). It has following benefits.

- Avoid unplanned shutdown due to breakdown and its consequential damage can be avoided
- Reduction in the failure rates and improving plant availability and reliability
- Reduction in spares inventory
- Reduction in unnecessary work due to sudden breakdown
- Possible efficient scheduling of repair/maintenance work to minimize downtime.

The CM should not be applied to all machines/structures/piping, etc. in a plant because this may not be cost-effective approach; hence, an optimized approach is used, which is illustrated in Figure 1.4. The selection of the items (machines or any

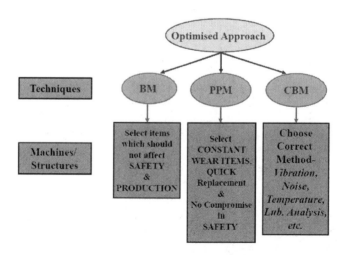

FIGURE 1.4 Viable approaches to minimize maintenance activities and cost.

Introduction

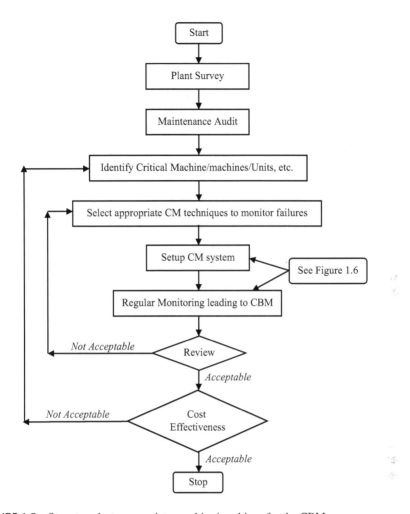

FIGURE 1.5 Steps to select appropriate machine/machines for the CBM.

other components) for the CM and CBM within a plant should follow the steps given in Figure 1.5. It is important to consider the following three elements for the selection of critical items.

1. Overall impact on the plant safety
2. High capital value means high repair cost in case of any failure
3. Potential production loss, hence not meeting the required demands

1.3.1 Lead-Time-to-Maintenance (LTM)

After the critical items are selected, the steps given in Figure 1.6 should be followed to the monitor the items and perform the maintenance activity if needed. The following, however, are still the biggest questions in the CBM practice.

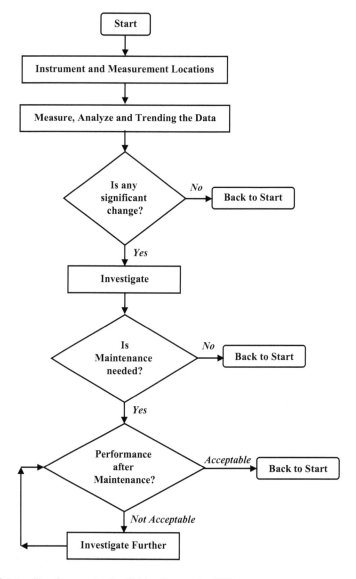

FIGURE 1.6 Simple steps in the CM leading to the CBM.

1. When is it best to carry out the maintenance work to rectify the defect?
2. Should it be immediately after identification of the defect by the CM system?
3. Can the maintenance personnel wait for some time to organize the maintenance activity without affecting the plant safety?

Therefore, it is always better to define a parameter, such as the Reliability and Maintenance Index (RMI), for each machine with their allowable and alarm limits

Introduction

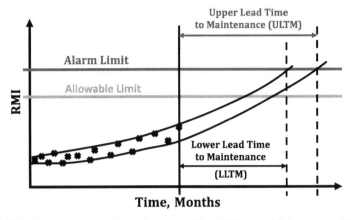

FIGURE 1.7 Practical approach to determine "Lead Time to Maintenance (LTM)" for a machine.

based on machine performance and CM data. Then it is good to trend this RMI on a daily or weekly basis, or at any pre-selected interval as shown in Figure 1.7. Industrial data for any machine may always show some scatter behavior; however, when this RMI is showing increasing trend, it is essential to extrapolate this trend data to predict the lead time to reach the alarm limit. This is also illustrated in Figure 1.7. This is termed as the Lead-Time-to-Maintenance (LTM). Two predictions are shown in Figure 1.7 i.e., Lower LTM (LLTM) and Upper LTM (ULTM) due to the scattered data. This simply means that the earliest maintenance can be planned by the LLTM with minimal risk of failure by this time. The ULTM is the possible maximum time that the maintenance personnel can wait for the maintenance activity, but this may have some of risk of failure.

Furthermore, it is good to have the predicated values of the LTM for all critical items within a plant. This may help to schedule a best possible period for the plant outage (or shutdown) to carry out all maintenance activities.

To aid the easy understanding for the estimation of the LLTM and ULTM, a rotating machine is considered. Examine the overall machine vibration velocity represents the RMI for the machine considered. The allowable and alarm limits for the vibration velocity are 1.5 and 2 mm/s respectively for this machine. The trend of the measured vibration values are shown in Figure 1.8. It is clear from the figure that there is no significant change in the vibration up to 60 days and then shows an increasing trend. The vibration measurements are carried out up to 170 days. The following equations can be used to estimate LTM.

$$V(t) = V_o \text{ when } 1 \leq t \leq t_o \tag{1.1}$$

$$V(t) = V_o e^{\lambda(t-t_o)} \text{ when } t \geq t_o \tag{1.2}$$

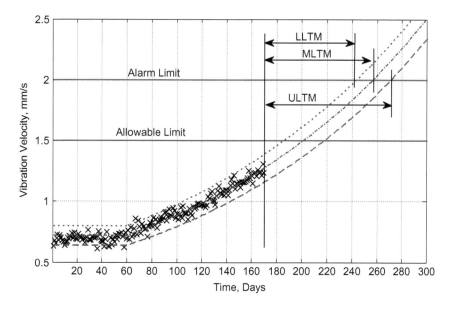

FIGURE 1.8 An example to estimate the LTM.

TABLE 1.1
List of Parameters Used and Estimated LTM Values

	V_o	t_o	λ	LTM
LLTM	0.80 mm/s	60 days	0.0050	73 days
ULTM	0.64 mm/s	60 days	0.0054	101 days
MLTM	0.072 mm/s	60 days	0.0052	86 days

where $V(t)$ is the RMI. V_o is the constant value of the RMI up to time, t_o and then it starts increasing. The value of λ defines the changing trend in the RMI values over the period of time.

For example consider, the following values are used to estimate the LLTM, ULTM and also mean LTM (MLTM). The selection of V_o and t_o values are taken from the vibration trend shown in Figure 1.8, which are listed in Table 1.1. The selection of λ is completely based on the trial and error approach to fit the vibration trend shown in Figure 1.8. The extrapolated curves for the LLTM, ULTM and MLTM are also shown in Figure 1.8; hence the LLTM, ULTM and MLTM values from the 170th day of measurement are estimated from the graph in Figure 1.8. These values also listed in Table 1.1.

Introduction 9

1.4 SUMMARY

The concepts of the CM and CBM are discussed briefly in the chapter. There are several NDT techniques can be used for the CM but the book is only focusing on the VCM of machines. The details of the instrumentation, measurement location, measurements and data analysis to identify the defects in rotating machines using the VCM are discussed through a number of chapters in this book.

2 Simple Vibration Theoretical Concept

2.1 EQUATION OF MOTION

To analyze a static problem, the usual practice is to balance the forces and moments acting on the static system. Similarly, for dynamic systems where the system status is changing with time, the vibrating system (or object) is assumed to be in *dynamic equilibrium* at a point of time by applying a fictitious force to bring the system to rest. This fictitious force is called an *inertia force*. The complete concept is known as the D'Alembert Principle, which is based on Newton's second law of motion.

In dynamics, the effect of the external disturbance other than gravity on the system is only studied. Hence, the dynamic behavior of a structure, in general, remains unaffected due to its any orientation in the space. For example, the behavior of a cantilever beam will always be same whether its orientation is vertical or horizontal or in an inclined plane.

The understanding of vibration theory is important for the professionals involved in vibration measurements and vibration-based condition monitoring (VCM). However, it is often difficult to write the dynamics equations for machine and structures, typically by many industrial professionals. But fully understanding the dynamics of a simple system such as a spring and mass system is always better. This concept can be utilized to understand the dynamics and vibration behavior of even a complex system.

A simple system consists of a spring of stiffness (k) and a mass (m) is considered here (Sinha, 2015). It is shown in Figure 2.1. This system is also known as a *single degree of freedom (SDOF) system* as the mass has an SDOF to oscillate in the y-direction only (along spring stiffness) as per Figure 2.1. While at rest the system is said to be in static equilibrium under gravity; however, when the system is disturbed from rest, it oscillates about the static equilibrium position along the y-direction. This oscillation is the natural oscillation of the mass within the spring-mass system without any external force. Hence the period of oscillation is call natural period and the related frequency as the natural frequency of this spring and mass system. Therefore the natural frequency is the property of this spring and mass system.

Mathematically, the equation of motion at a time, t, for the SDOF system shown in Figure 2.1 is written as

$$F_i(t) + F_s(t) = 0 \qquad (2.1)$$

where $F_i(t)$ and $F_s(t)$ are the inertia force and the stiffness force at a time, t respectively. Hence the inertia force due to the oscillation of the mass is completely

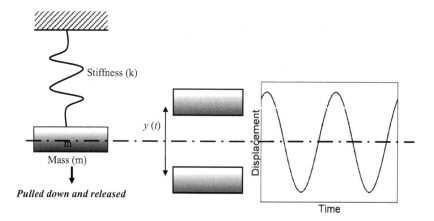

FIGURE 2.1 An SDOF system and its behavior during free vibration.

balanced by the stiffness force within the spring-mass system without any external force during the natural oscillation of the mass. So Equation (2.1) becomes

$$m\ddot{y}(t) + ky(t) = 0 \tag{2.2}$$

It is known that the system undergoes cyclic (harmonic) motion, therefore Equation (2.2) may be written as

$$-m\omega_n^2 y(t) + ky(t) = 0 \tag{2.3}$$

where $\omega_n = \sqrt{\frac{k}{m}}$ is known as the Circular (or Angular) frequency. So, the period of oscillation is

$$T = \frac{2\pi}{\omega_n} \tag{2.4}$$

The frequency related to Time period, T, is

$$f_n = \frac{1}{T} = \frac{\omega_n}{2\pi} = \frac{1}{2\pi}\sqrt{\frac{k}{m}} \tag{2.5}$$

where f_n is called the *Natural Frequency* of the spring-mass system; as it depends on the properties (k, m) of the system. The unit of frequency is *Cycle/s* or *Hz* (Hertz).

The general solution of Equation (2.3), which is a homogeneous second order linear differential equation, is given by

$$y(t) = A \sin \omega_n t + B \cos \omega_n t \tag{2.6}$$

Simple Vibration Theoretical Concept

where the constants A and B are evaluated from initial conditions; $y(t=0) = y(0)$, and $\dot{y}(t=0) = \dot{y}(0)$. Therefore, Equation (2.6) become

$$y(t) = \frac{\dot{y}(0)}{\omega_n}\sin\omega_n t + y(0)\cos\omega_n t \tag{2.7}$$

2.2 DAMPED SYSTEM

In practice, none of the real systems can freely oscillate with the same amplitude of vibration for an infinite time without the aid of any external source of excitation (see Figure 2.2). The natural tendency of any system is to decay down to the equilibrium position after a few oscillations when the system is disturbed from its natural equilibrium position. This phenomenon simply indicates that there is some kind of energy dissipation during oscillation. This inherent tendency is classified as damping of the system in the domain of structural dynamics. Hence, the equivalent dynamic system of a spring and mass SDOF system is often represented as shown in Figure 2.3a; where the dash-pot is used to represent the damping in the system. Figure 2.3b typically shows the decay motion during the free vibration of the system.

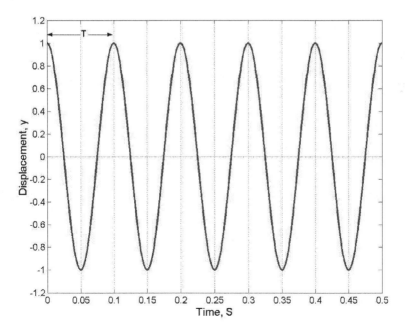

FIGURE 2.2 Response of an SDOF system with the natural period of oscillation (T) equals to 0.1s.

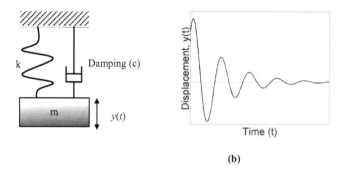

FIGURE 2.3 A damped SDOF system and its behavior. (a) Representation of a dynamic system. (b) Free vibration.

2.2.1 Equation of Motion for Free Vibration

The equation of motion (Equation (2.1)) for the damped system undergoing free vibration when disturbed from an initial equilibrium position at a time, t, is now written as

$$F_i(t) + F_c(t) + F_s(t) = 0 \tag{2.8}$$

$\Rightarrow$ Inertia Force + Damping Force + Stiffness Force = 0

where the additional term, $F_c(t)$, is called the damping force which is assumed to be dependent on the velocity of the vibration, and is defined as $F_c(t) = c\dot{y}(t)$, where c is the damping constant of the system. Hence Equation (2.8) can be written as

$$m\ddot{y}(t) + c\dot{y}(t) + ky(t) = 0 \tag{2.9}$$

Let us assume the solution of Equation (2.9) in the form

$$y(t) = y_0 e^{st} \tag{2.10}$$

where y_0 and s are constants. Upon substitution, the Equation (2.9) becomes

$$(ms^2 + cs + k)y_0 e^{st} = 0 \tag{2.11}$$

Equation (2.11) will be satisfied for all values of t when

$$ms^2 + cs + k = 0$$
$$\Rightarrow s^2 + \frac{c}{m}s + \frac{k}{m} = 0 \tag{2.12}$$

Equation (2.12) has two roots

$$s_{1,2} = -\frac{c}{2m} \pm \sqrt{\left(\frac{c}{2m}\right)^2 - \frac{k}{m}} \tag{2.13}$$

Simple Vibration Theoretical Concept   **15**

Hence the general solution of Equation (2.10) is given by

$$y(t) = Ae^{s_1 t} + Be^{s_2 t} \qquad (2.14)$$

where A and B are constants which are estimated from the initial conditions, $y(t = 0) = y(0)$, and $\dot{y}(t = 0) = \dot{y}(0)$ of the system. However the system behavior is completely dependent on the Discriminant factor (DF) $= \sqrt{(\frac{c}{2m})^2 - \frac{k}{m}}$ of the roots s_1 and s_2. All three possibilities are discussed here.

2.2.2 CRITICALLY DAMPED SYSTEM

Case (I) Limiting Case When DF = 0

The expression, $\sqrt{(\frac{c}{2m})^2 - \frac{k}{m}} = 0$ gives $c = 2m\sqrt{\frac{k}{m}} = 2m\omega_n = 2\sqrt{km}$. This damping is called *critical damping*, c_c. For this case, both roots are equal; hence, solution (2.14) becomes

$$y(t) = (A + Bt)e^{st} \qquad (2.15)$$

where $s = s_1 = s_2 = -\frac{c}{2m}$. In practice the damping (c) of the system is expressed in terms of the critical damping (c_c) by a non-dimensional number (ζ), called the *damping ratio*. The root, s, is now written as

$$s = -\frac{c}{2m} = -\frac{c}{c_c}\frac{c_c}{2m} = -\zeta\frac{2m\omega_n}{2m} = -\zeta\omega_n \qquad (2.16)$$

Here, the damping, c, is equal to the critical damping, c_c, i.e., $\zeta = 1$, and the root, $s = -\omega_n$. Upon substitution, Equation (2.15) is written as

$$y(t) = (A + Bt)e^{-\omega_n t} \qquad (2.17)$$

Substitution of initial conditions $(y(0)$ and $\dot{y}(0))$ in Equation (2.17) gives

$$y(t) = \left(y(0) + \left(\dot{y}(0) + \omega_n y(0)\right)t\right)e^{-\omega_n t} \qquad (2.18)$$

With damping ratio, $\zeta = 1$, the response, $y(t)$, of the system is shown in Figure 2.4. There is no oscillatory motion in the system response if the system is disturbed from the equilibrium position; hence, this condition is known as a *critically damped system* and is the limiting case between the oscillatory and non-oscillatory motion. This simply means that the system will return back to its equilibrium position in minimum possible time without undergoing any oscillatory motion when disturbed from the equilibrium position.

2.2.3 OVER DAMPED SYSTEM

Case (II) When DF > 0

The expression, $\sqrt{(\frac{c}{2m})^2 - \frac{k}{m}} > 0$ gives, $c > c_c$, and $\zeta > 1$. This system is called as the *over damped system* where the two roots, s_1 and s_2, remain real. Since the damping ratio is more than 1, the motion will be non-oscillatory. The general solution in Equation (2.14) now becomes

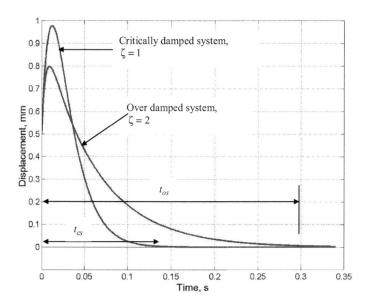

FIGURE 2.4 Comparison of dynamic behavior between an over damped and a critically damped systems.

$$y(t) = Ae^{\left(-\zeta+\sqrt{\zeta^2-1}\right)\omega_n t} + Be^{\left(-\zeta-\sqrt{\zeta^2-1}\right)\omega_n t} \quad (2.19)$$

where

$$A = \frac{\dot{y}(0)+\left(\zeta+\sqrt{\zeta^2-1}\right)\omega_n y(0)}{2\omega_n\sqrt{\zeta^2-1}}, \text{ and } B = \frac{-\dot{y}(0)-\left(\zeta-\sqrt{\zeta^2-1}\right)\omega_n y(0)}{2\omega_n\sqrt{\zeta^2-1}} \quad (2.20)$$

Figure 2.4 also illustrates the behavior of an over damped system, $\zeta = 2$ compared to a critically damped system, $\zeta = 1$, but the same system and same initial conditions are used for both cases. It is obvious from the responses shown in Figure 2.4 that the over damped system requires a larger time ($t_{os} > t_{cs}$) to return to the equilibrium position compared to the critically damped system.

2.2.4 Under Damped System

Case (III) When DF < 0
For this case, the expression, $\sqrt{(\frac{c}{2m})^2 - \frac{k}{m}} < 0$, gives, $c < c_c$, and $\zeta < 1$. This is the case of an *under damped system*, and the system will have an oscillatory motion; hence, this case represents most of the real life systems, but both roots do not remain real. The roots (Equation (2.13)) are written as

$$s_{1,2} = -\frac{c}{2m} \pm j\sqrt{\frac{k}{m}-\left(\frac{c}{2m}\right)^2} \quad (2.21)$$

Simple Vibration Theoretical Concept

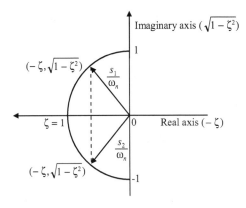

FIGURE 2.5 Representation of poles.

$$\Rightarrow s_{1,2} = -\zeta\omega_n \pm j\sqrt{1-\zeta^2}\,\omega_n \qquad (2.22)$$

where $j = \sqrt{-1}$ is an imaginary number. The roots are now complex conjugates. They are known as *poles* of the system in structural dynamics and are often represented in the real and imaginary axes as shown in Figure 2.5.

The roots (*poles*) are also expressed as

$$s_{1,2} = \lambda_{1,2} = \pm j\sqrt{1-\zeta^2}\,\omega_n = -\zeta\omega_n \pm j\omega_d \qquad (2.23)$$

where $\omega_d = \omega_n\sqrt{1-\zeta^2}$ is the *damped natural frequency* of the SDOF system. The general solution (Equation (2.14)) becomes

$$y(t) = e^{-\zeta\omega_n t}\left(Ae^{j\sqrt{1-\zeta^2}\,\omega_n t} + Be^{-j\sqrt{1-\zeta^2}\,\omega_n t}\right) \qquad (2.24)$$

$$\Rightarrow y(t) = e^{-\zeta\omega_n t}\left(Ae^{j\omega_d t} + Be^{-j\omega_d t}\right) \qquad (2.25)$$

Equation (2.24) can further be written as

$$y(t) = e^{-\zeta\omega_n t}(C_1 \cos\omega_d t + C_2 \sin\omega_d t) \qquad (2.26)$$

where $C_1 = y(0)$, $C_2 = \frac{\dot{y}(0) + \zeta\omega_n y(0)}{\omega_d}$. Equation (2.26) can also be written as

$$y(t) = Ye^{-\zeta\omega_n t}\sin(\omega_d t + \theta) \qquad (2.27)$$

where $Y = \sqrt{C_1^2 + C_2^2}$, and the angle, $\theta = \tan^{-1}\left(\frac{C_1}{C_2}\right)$.

Figure 2.6 typically illustrates the dynamic behavior of the *under damped system* when disturbed from the equilibrium position.

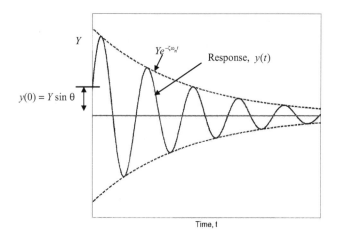

FIGURE 2.6 Response of the under damped SDOF system.

2.3 FORCED VIBRATION

When a system vibrates under the influence of an external force, this is called forced vibration. It is schematically shown in Figure 2.7. The equation of motion in Equation (2.8) is now modified as

$$F_i(t) + F_c(t) + F_s(t) = F(t) \tag{2.28}$$

$$\Rightarrow m\ddot{y}(t) + c\dot{y}(t) + ky(t) = F(t) \tag{2.29}$$

Hence, Equation (2.29) will have two solutions;

1. *Complimentary solution*, $y_c(t)$, when

$$m\ddot{y}(t) + c\dot{y}(t) + ky(t) = 0$$

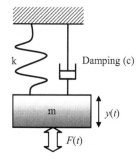

FIGURE 2.7 An under damped SDOF system vibrating under an external force.

Simple Vibration Theoretical Concept

It is nothing but the response of the system under free vibration, and the solution, $y_c(t)$, is Equation (2.26) or (2.27).

2. *Particular Integration due to the applied force*, $y_p(t)$. So the complete solution is given by

$$y(t) = y_c(t) + y_p(t) \tag{2.30}$$

The complete response of the system is the combined effect of the free vibration and the response due to the externally applied force. The response $(y_c(t))$ is called as the *transient response*, whereas the response $(y_p(t))$ is the *steady state response*.

For *Particular Integration*, the general solution is assumed to be $y_p(t) = Ye^{j\omega t}$ for the applied force, $F(t) = F_0 e^{j\omega t}$, where ω is the forcing frequency in rad./s. Upon substitution, Equation (2.29) becomes

$$(-m\omega^2 + jc\omega + k)Ye^{j\omega t} = F_0 e^{j\omega t} \tag{2.31}$$

Hence, the displacement, $y_p(t)$, is given by

$$y_p(t) = \frac{F_0 e^{j\omega t}}{(k - m\omega^2) + jc\omega} \tag{2.32}$$

$$\Rightarrow y_p(t) = \frac{(F_0 / k)}{\left[1 - \left(\dfrac{\omega}{\omega_n}\right)^2\right] + j\left(2\zeta\dfrac{\omega}{\omega_n}\right)} e^{j\omega t} \tag{2.33}$$

Equation (2.33) is a complex quantity. Any complex quantity $(a + jb)$ can be expressed as $Ae^{j\theta}$, where $A = \sqrt{a^2 + b^2}$, and $\theta = \tan^{-1}\left(\frac{b}{a}\right)$. Hence, Equation (2.33) is expressed as

$$y_p(t) = \frac{(F_0 / k)e^{j\omega t}}{\sqrt{\left[1 - \left(\dfrac{\omega}{\omega_n}\right)^2\right]^2 + \left(2\zeta\dfrac{\omega}{\omega_n}\right)^2} e^{j\phi}} \tag{2.34}$$

$$\Rightarrow y_p(t) = \frac{(F_0 / k)}{\sqrt{\left(1 - r^2\right)^2 + \left(2\zeta r\right)^2}} e^{j(\omega t - \phi)} \tag{2.35}$$

$$\Rightarrow y_p(t) = Ye^{j(\omega t - \phi)} \tag{2.36}$$

where $r = \frac{f}{f_n} = \frac{\omega}{\omega_n}$, Y is the maximum steady state response (displacement), and ϕ is the phase of the response with respect to the exciting force. The phase angle is given by

$$\phi = \tan^{-1}\frac{2\zeta r}{(1 - r^2)} \tag{2.37}$$

Equation (2.35) can further be written as

$$y_p(t) = \left(\frac{F_0}{k}\right) D e^{j(\omega t - \phi)} \tag{2.38}$$

where $\left(\frac{F_0}{k}\right)$ is equivalent to the static deflection of the system due to applied load of amplitude, F_0, and $D = \frac{1}{\sqrt{(1-r^2)^2 + (2\zeta r)^2}}$ is called an *Amplification* or *Magnification Factor*. When the forcing frequency $(\omega = 2\pi f)$ becomes equal to the system natural frequency $(\omega_n = 2\pi f_n)$, the phenomenon is called resonance; the amplification in the response takes place during each cycle of the oscillations of the system. The factor for the maximum possible amplification in the system response at resonance, is given by

$$D = \frac{1}{2\zeta}, \text{ when } r = 1 \tag{2.39}$$

Hence, if a system with damping, $\zeta = 0$, then $D \to \infty$, $y_p(t) \to \infty$, $y(t) \to \infty$, over a period of oscillation irrespective of the amplitude of the exciting force and the system natural frequency. The build-up of such a vibration response at resonance for a spring-mass system with natural frequency, $f_n = 10$ Hz but the damping, $\zeta = 0$, and applied force, $F(t) = \sin \omega_n t$, is shown in Figure 2.8. It can be seen from Figure 2.8 that the response is linearly increasing with each cycle of oscillation and will reach to infinity over a period of time. If the amplitude of the applied force is larger, then the system will reach an infinite displacement in a shorter time.

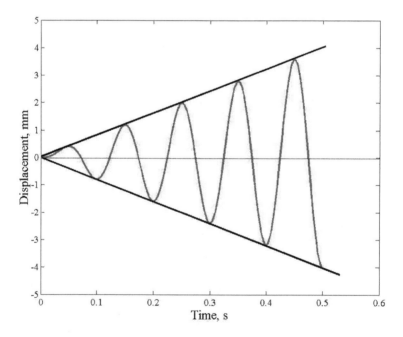

FIGURE 2.8 Response built-up of a system at the resonance when $\zeta = 0$.

Simple Vibration Theoretical Concept

The build-up of the vibration response of the same system with the same applied force but the damping, $\zeta = 8\%$ is shown in Figure 2.9 for comparison. The displacement response is calculated using Equation (2.30). The transient and steady state responses are clearly shown in Figure 2.9.

Figure 2.10 represents the change in the vibration amplitude (D) and phase of the vibration response to the applied force for an SDOF (spring-mass system) with applied frequencies with different system damping ratios. The equations of the amplification factor (D) and the phase angle (Equation (2.37)) are used to generate

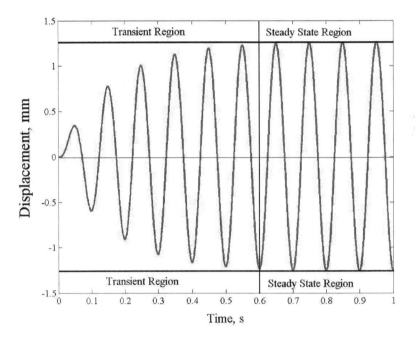

FIGURE 2.9 Response built-up of a system at the resonance when $\zeta = 8\%$.

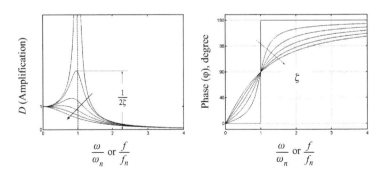

FIGURE 2.10 Steady state amplification and phase relation between displacement and force at different exciting frequencies.

22   Industrial Approaches in Vibration-Based Condition Monitoring

Figure 2.10. The frequency ratio (forcing frequency to the natural frequency), $r = 1$ always represents the resonance phenomenon where the response amplitude can go up to "infinity" depending upon the system damping and phase between the response and force is always equal to 90°. This is also evident from Figure 2.10 that the change in the vibration response phase angle is equal to 180° with respect to the applied force if the forcing frequency is changing from the low frequency to high frequency passing through 90° are the resonance.

2.3.1 Example 2.1: An SDOF System

It is consisting of a mass, $m = 1.0132$ kg, a spring of stiffness, $k = 1000$ N/m with the damping ratio of 0.1% is considered here. Now it is assumed that a force, $(t) = F_0 \sin(2\pi ft)$, where $F_0 = 1$ N is applied to the mass but at different frequencies, 0 Hz (i.e., static force, $F_0 = 1$ N), 0.5 Hz (i.e., 1 N cyclic load at 30 Cycle per minute, CPM), 2.5 Hz (150 CPM), 3.5 Hz (210 CPM), 5 Hz (300 CPM), 6.5 Hz (390 CPM) and 7.5 Hz (450 CPM) step-by-step.

The natural frequency of this system is estimated as 5 Hz using Equation (2.5). Now the maximum response of the system is then calculated for each forcing frequency using Equation (2.35). The calculated maximum responses are tabulated in Table 2.1. It is clear from Table 2.1 that the vibration amplitudes are completely different ranging from 0.80 to 500 mm for the same applied force, $F_0 = 1$ N but due to different forcing frequencies. When applied frequency is equal to or close to the system natural frequency then the vibration response is very high which has potential to damage the system in few numbers of oscillations. The vibration responses of the system at different frequencies are shown in Figure 2.11.

Let's consider now the system in Example 2.1 is nothing but a rotating machine. If the machine rotating speed (rotation per minute, RPM) is close to or equal to 300 CPM (5 Hz—system natural frequency) then the machine is likely to fail

TABLE 2.1

Calculated Maximum Responses for the System in Example 2.1

Case	Forcing Frequency, f	System: $m = 1.0132$ kg, $k = 1$ N/mm, $\zeta = 0.001$, $f_n = 5$ Hz		
		Frequency ratio, r	Max. displ., mm	Remarks
	0 Hz	0	1.00	No vibration i.e., static load of 1 N and the displacement is just static deflection
	0.5 Hz	0.1	1.01	Displacement increased due to vibratory force at 0.5 Hz
	2.5 Hz	0.5	1.33	Further increased at $f = 2.5$ Hz
	3.5 Hz	0.7	1.96	Further increased at $f = 3.5$ Hz
	5 Hz	1.0	500	Maximum due resonance
	6.5 Hz	1.3	1.45	Displacement starts decreasing beyond resonance
	7.5 Hz	1.5	0.80	

Simple Vibration Theoretical Concept

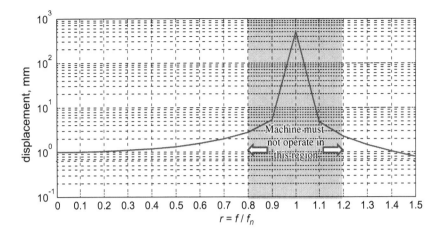

FIGURE 2.11 Displacement responses at different forcing frequencies for Example 2.1.

frequently even if the design of the machine is perfectly fine. Therefore it is always important to avoid such situation for any rotating machine by shifting the machine natural frequency during installation. It is marked in Figure 2.11. This figure is showing ±20% away from the natural frequency. This may be acceptable for this system. But the ±20% range should not be used as the thumb rule for all systems. It may be different for different systems. It is difficult to recommend any specific value. The frequency response function plots from the *in-situ* modal tests (Chapter 8) on the machine can provide some estimate on this i.e., how much away from the natural frequency is acceptable for the safe machine operation.

2.4 CONCEPT OF MODESHAPES

The mass in the spring-mass system will have one complete oscillation from "zero-to-maximum deflection-to-zero-to-minimum deflection-to-zero" during a natural time period when vibrating at its natural frequency. However most of the real systems are complex and continuous, and may not like a spring-mass system. So most of the real systems will have an infinite number of natural frequencies, i.e., different combinations of masses and springs within the real systems (structures or machines) generally yield many natural frequencies. Hence the expression given in Equation (2.5) is only valid for a spring-mass system only. But still this concept (no exact formula) can be used to explain each natural frequency of any real systems.

When any system is vibrating at a natural frequency then the each point within the system is going to oscillate sinusoidally at that particular natural frequency but the amplitude and direction of deformation of each point may or may not be same. This typical shape of deformation of the system at a natural frequency is called modeshape for that particular natural frequency.

To aid the better understanding, a cantilever beam is chosen here to explain the concept of the modeshapes. The left end of the beam is assumed to be clamped

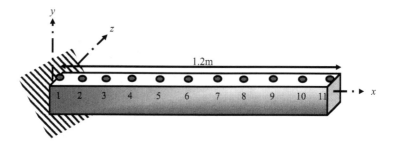

FIGURE 2.12 A cantilever beam.

to satisfy the condition of the cantilever beam as shown in Figure 2.12. Now consider the beam has 11 locations for the vibration measurements in vertical direction (y-direction). The measured first four natural frequencies in the vertical direction (i.e., bending modes) are 11.64 Hz, 72.96 Hz, 204.33 Hz, and 400.69 Hz, respectively. The measured corresponding modeshapes are listed in Table 2.2. The concept of measuring natural frequencies and their modeshapes for any systems is explained in Chapter 8.

The modeshapes related to the bending deflection in Table 2.2 for Modes 1–4 are shown in Figure 2.13 that represents the deformation pattern at each mode. To further explain the concept of the modeshape deformation pattern, the beam is assumed to be vibrating at the first mode, $f_{n_1} = 11.64$ Hz where the time period, $T1 = 1/f_{n_1} = 0.068$ s $= 68$ ms. This means that each point on the beam is undergoing through a complete sinusoidal cycle in this time period, T1 as per the 1st modeshape. It is also explained pictorially in Figure 2.14. Figures on the left-hand side column show the vibration of

TABLE 2.2
Measured Modeshapes Values

Measurement Location No	Location, m	$\varphi_{y,1}$	$\varphi_{y,2}$	$\varphi_{y,3}$	$\varphi_{y,4}$
1	0	0	0	0	0
2	0.12	1.7338e-02	9.5750e-02	−2.3586e-01	−3.9871e-01
3	0.24	6.6019e-02	3.1120e-01	−6.2515e-01	−7.8063e-01
4	0.36	1.4107e-01	5.4386e-01	−7.8207e-01	−4.4936e-01
5	0.48	2.3761e-01	7.0650e-01	−5.4389e-01	3.2674e-01
6	0.60	3.5094e-01	7.3771e-01	−2.0370e-02	7.3230e-01
7	0.72	4.7664e-01	6.0934e-01	4.8994e-01	3.3819e-01
8	0.84	6.1074e-01	3.2773e-01	6.7988e-01	−4.1146e-01
9	0.96	7.4987e-01	−7.2395e-02	4.0837e-01	−6.6594e-01
10	1.08	8.9140e-01	−5.4140e-01	−2.3630e-01	−5.3982e-02
11	1.20	1.0336e+00	−1.0337e+00	−1.0341e+00	1.0355e+00
Natural frequencies		$f_{n_1} = 11.64$Hz	$f_{n_2} = 72.96$Hz	$f_{n_3} = 204.33$Hz	$f_{n_4} = 400.69$Hz

Simple Vibration Theoretical Concept

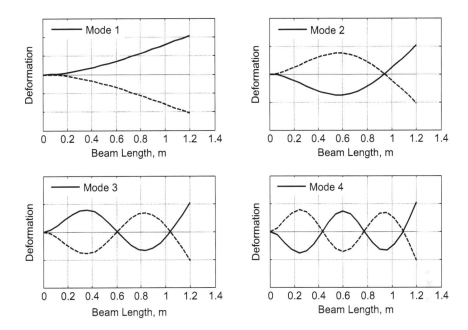

FIGURE 2.13 Modeshapes of the cantilever beam.

the measured location 11 (free end) with a circle pointer whereas the figures on the right-hand side column show the beam deformation corresponding to the instant of the circle pointer during the beam vibration. "Arrow heads" in Figure 2.14 show the direction of beam motion at specific instant of vibration of the beam.

Similarly the mode 2, $f_{n2} = 72.96$ Hz and the time period, T2 = 13.70 ms, the shape of deformation with time is shown in Figure 2.15. It is clear that the measurement locations 1–8 are moving in phase with different amplitudes. The remaining locations 9–11 are also moving in phase but out of phase to the locations 1–8.

2.5 MACHINE VIBRATION

A horizontal rotating rig (Sinha, 2002) is shown in Figure 2.16. The rotor is carrying two-balance discs supported through Bearings A and B, which are mounted on separate flexible foundations. The rig is driven by a motor. The motor shaft and the rig rotor are connected through the flexible coupling so that vibration of motor part is completely ignored here.

2.5.1 Rotor Dynamics

During the machine operation (i.e., shaft rotation) the rotor likely to be deformation in both vertical and horizontal directions for the horizontal rig in Figure 2.16. Therefore it is normal practice to show the modeshapes at different modes in both axes simultaneously to trace the path of rotor deformation during the machine vibration. Few modeshapes of the rig are typically shown in Figure 2.17.

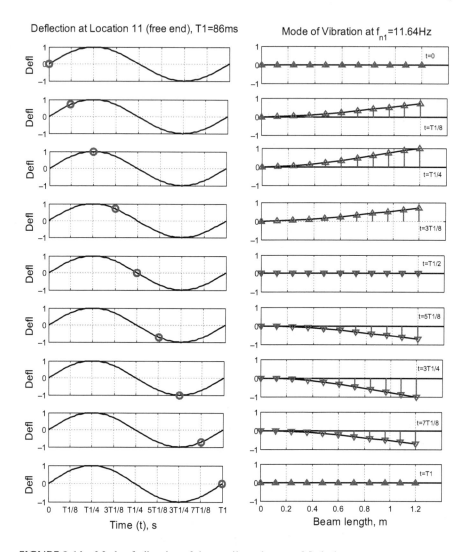

FIGURE 2.14 Mode of vibration of the cantilever beam at Mode 1.

Figure 2.18 typically shows the vibration responses at Bearing A of the rig in the horizontal direction when the rig is running down linearly from the rotating speed of 2500–300 RPM. This clearly indicates that the rig is vibrating and the vibration acceleration amplitudes are generally showing decreasing trend if the amplification parts of vibration are ignored.

Therefore it is important to understand the following questions:

1. Why does the machine vibrate when operating?
2. Why do the vibration amplitudes show a decreasing trend in general from high rotating speed to low rotating speed?
3. Why are there a few vibration amplifications during machine rundown?

Simple Vibration Theoretical Concept

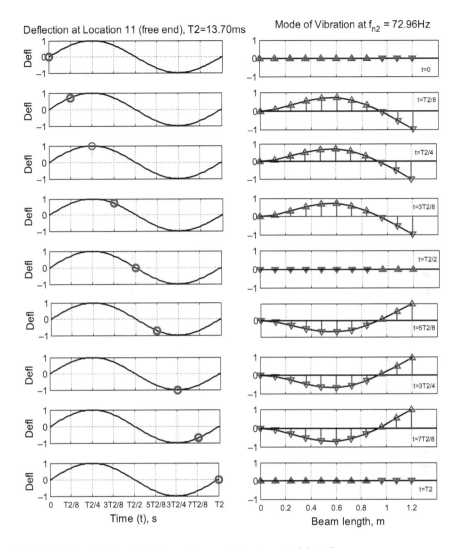

FIGURE 2.15 Mode of vibration of the cantilever beam at Mode 2.

2.5.2 Unbalance Responses

The primary reason for the machine vibration is due to the rotor unbalance force. The rotor unbalance force is originated from the concept of the centrifugal force acting on an object during circular motion. This is demonstrated through a simple example shown in Figure 2.19. Figure 2.19 is showing a stone of mass (m) connected through a thread of length (r), which is rotating in the horizontal plane with a tangential velocity (v). Therefore a radial force acting on the mass away from the center of rotation at any time, t is known as the centrifugal force, which is equal to $F_{Centrifugal} = \frac{mv^2}{r} = mr\omega^2$, where $v = \omega r$, ω is the angular velocity, radian/s (rad/s).

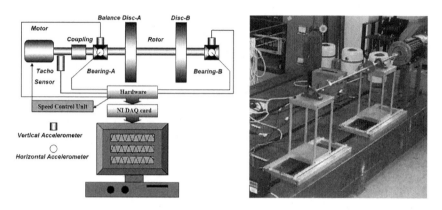

FIGURE 2.16 An experimental rig with vibration instruments—schematic and photograph.

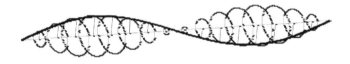

FIGURE 2.17 Modeshapes of a rotor.

Simple Vibration Theoretical Concept

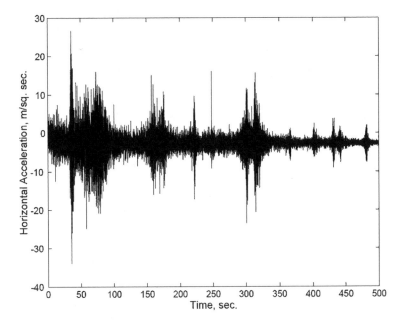

FIGURE 2.18 Vibration acceleration at Bearing A in horizontal direction.

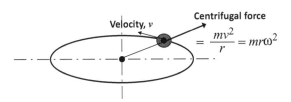

FIGURE 2.19 A concept of the centrifugal force.

A rotor with a balance disc is also shown in Figure 2.20. It is assumed that the shaft is weightless for the clarity only but the balance disc has a resultant unbalance mass (m) at a radius (r). The possible reason for this unbalance mass even in new condition is that it is often difficult to have any material with uniform density distribution and keeping uniform dimension during machining process. Hence the centrifugal force of amplitude, $F_{Centrifugal} = F_{Unbalance} = mr\omega^2$ is likely to acting on the disc if the rotor is rotating at speed, $f = 50$ rotation per s or 3000 RPM (or $\omega = 2\pi f$ rad/s). This centrifugal force in the rotation machine is called the rotor unbalance force. This is always constant in the radial direction at any time of the shaft rotation. Although it is a constant radial force but it generates vibration in the machine.

To aid the simple understanding of vibration generation in any machine, the cross-section of a rotor with a bearing is consider here which is shown in Figure 2.21. It is

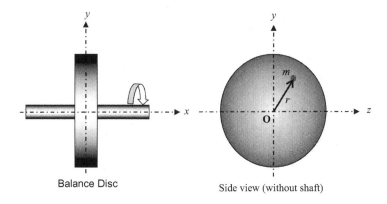

FIGURE 2.20 A rotor with a single plane unbalance.

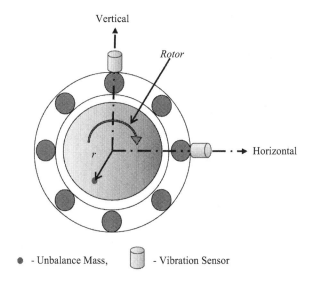

FIGURE 2.21 Measurement of machine vibration on bearing housing.

also assumed that the vibration is measured in both vertical and horizontal directions simultaneously. A solid point on the rotor is representing the unbalance mass location on the rotor; hence, a constant unbalance force will always be acting in the radial direction due to this unbalance mass. The time for one complete rotor rotation (360°) will be equal to 1/50 = 0.02 s.

The mechanism of vibration generation due this constant radial unbalance force is shown in Figure 2.22. At Location 1 (time $t = 0$ s, angle = 0°), the unbalance mass location is at top position so that the unbalance force is acting in positive vertical direction and hence generates "+ve maximum vibration" in the vertical direction and "zero vibration" in the horizontal direction. Similarly it is illustrated for Locations 2

Simple Vibration Theoretical Concept

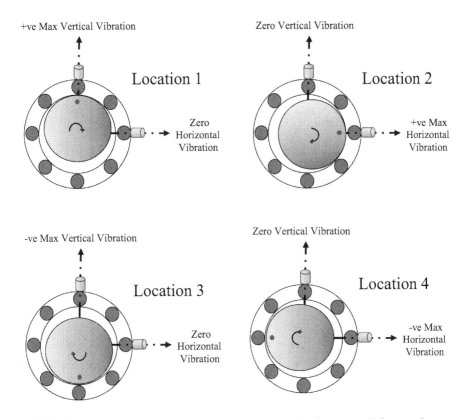

FIGURE 2.22 Typical machine vibration mechanisms under the rotor unbalance only.

(time $t = 0.005$ s, angle $= 90°$), 3 (time $t = 0.01$ s, angle $= 180°$) and 4 (time $t = 0.015$ s, angle $= 270°$).

It is clear from this simple illustration that the each point in a machine is subjected to a sinusoidal force due to the constant unbalance force acting in the radial direction. Hence the sinusoidal response is expected for any healthy machine. If the amplitude of the sinusoidal response is high then it is recommended to do the rotor balancing to reduce the rotor unbalance. Typical vibration responses in the vertical and horizontal directions under the rotor unbalance force for the illustrated example in Figure 2.22 are shown in Figure 2.23. The responses are pure sine waves for both vertical and horizontal directions but at the phase of 90°. Therefore the SDOF equation of motion in Equation (2.29) can also represents a machine vibration in a direction (vertical or horizontal or radial) which can be written as

$$m\ddot{y}_r(t) + c\dot{y}_r(t) + ky_r(t) = F_{Unbalance}\sin(\omega t + \phi) \quad (2.40)$$

where $y_r(t)$ is the radial displacement and ϕ is the phase angle of the unbalance mass with respect to a stationary reference in the machine. Equation (2.40)

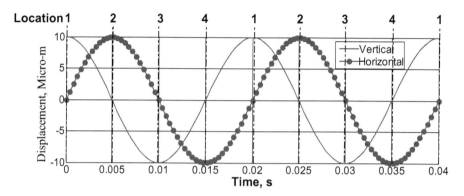

FIGURE 2.23 Typical sinusoidal vibration responses of machine due to the rotor unbalance only.

is nothing but representing Jeffcott Rotor model without the gyroscopic effect. It is equivalent to a simple SDOF system. The gyroscopic effect is deliberately not included here to keep it simple for the vibration-based condition monitoring (VCM) professionals. They can always refer other books on rotor dynamics for further details on this.

The amplitude of the unbalance force, $F_{Unbalance} = mr\omega^2$ is dependent on the rotating speed. If the speed is high then the force will be higher and vice versa. This is the reason why amplitude of vibration often tends to shown decreasing amplitude from high speed to low speed of machine.

Generally the amplification of vibration during machine runup or rundown indicates that the machine speed is passing through the machine natural frequency. This machine natural frequency is known as the critical speed of the machine. Many machines may have several natural frequencies (or critical speeds) in the speed range of zero speed to the machine operating speed.

2.5.3 Machine Faults

If there is any defect developed in any machine during machine operation such as mechanical looseness, rotor crack, etc. then Equation (2.40) is likely to change. The defect may generate a defect related force in addition to the existing unbalance force and/or time dependent change in the system stiffness; hence, the vibration response is not likely to be pure sinusoidal response. It may change to periodic response with the fundamental frequency related to the machine RPM. A typical response is shown in Figure 2.24a.

Similarly a defect (either inner race or out race or ball) in a ball bearing is going to generates an impulsive signal during each shaft rotation; therefore, the vibration response of machine is likely to contain both rotor unbalance response plus response due to bearing defect. A typical signal is shown in Figure 2.24b.

Simple Vibration Theoretical Concept

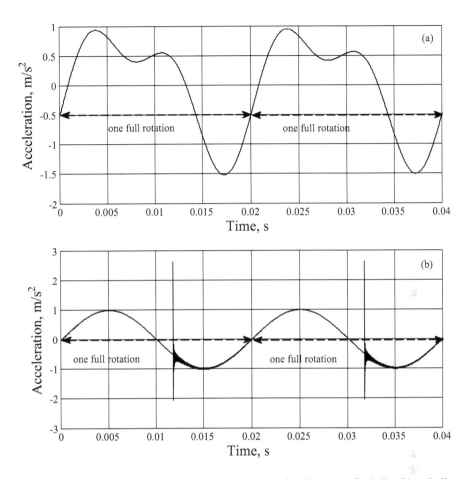

FIGURE 2.24 Vibration responses for machine having fault/faults: (a) Periodic, (b) periodic plus impulsive.

2.6 SUMMARY

A very simple concept of vibration theory using a spring-mass system is explained in the chapter. The theoretical concept is then extended to the structures and machines vibration without any complicated mathematics. The mechanism of machine vibration is also introduced in a very simple way.

REFERENCES

Sinha, Jyoti K. 2002. Health Monitoring Techniques for Rotating Machinery, PhD Thesis, University of Wales Swansea (Swansea University), Swansea, UK, October 2002.

Sinha, Jyoti K. 2015. *Vibration Analysis, Instruments, and Signal Processing.* Boca Raton, FL: CRC Press/Taylor & Francis Group, January 2015. http://www.crcpress.com/product/isbn/9781482231441.

3 Vibration-Based Condition Monitoring and Fault Diagnosis
Step-by-Step Approach

3.1 INTRODUCTION

The VCM within industries is generally known as a tool to identify the fault(s) so that the identified fault can be fixed during the planned shutdown. The VCM role, however, is much broader that includes investigating the reason for the frequent and repetitive failures in any machines by doing either additional vibration tests or further analysis of the measured vibration data in addition to the regular machine condition monitoring. The chapter provides the basic concept on the industrial vibration-based condition monitoring (VCM) for any rotating machines.

3.2 DIFFERENT STAGES OF VIBRATION MEASUREMENTS AND MONITORING

3.2.1 BATH TUB CONCEPT

Figure 3.1 represents a typical bath tub concept used in industries. It has three phases, namely, *Early failure*, *Useful life*, and *Old age*. This bath tub concept is nearly true for most machines and structures. This represents the life cycle model for any object but the time span of each stage can be different even for the identical machines and structures. It is also difficult to estimate the life even if factors such as loading and environment are known and stay constant. In real applications these factors may not be known fully which makes the life prediction process even more complex. Therefore the CM should be applied during the complete life cycle definitely for critical machines. For simplification, the VCM approach for any machines is divided into three stages using the bath tub concept.

3.2.2 STAGE 1—MACHINE INSTALLATION AND COMMISSIONING

For machines, the early failure may not be related to their poor design in most cases. In fact the exactly identical machines may be working perfectly fine within the same plant or in many other industries. The reason for the failure

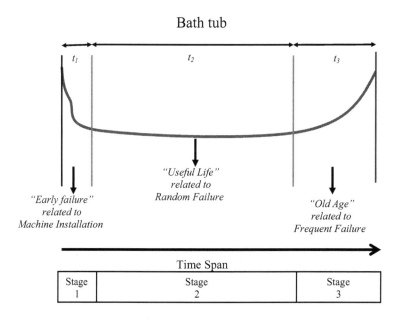

FIGURE 3.1 A simple bath tab concept representing life cycle model.

may be different. Machine generally consists of three parts- rotor, bearing and foundations including all auxiliary components. An abstract view of a machine is shown in Figure 3.2. The assembly and foundation of any machine during installation generally plays significant role in the early failure. It is often observed that a number of identical machines (say motor-pump units) are used in a plant but their installation foundation may not be identical either due to space limitation or due to easy and convenient approach used during installation. It is also frequently observed that only few machines are having frequent problems out of all identical machines.

The reason for such observation is not because of the fault in the design of those particular machines in most cases but it may be due their different dynamics behavior resulting from their foundations. It is because dynamics of any machine is the combined effect of the rotor, bearing and foundation. In most cases it has been observed that the one of the machine natural frequencies of the as-installed machine is close to the machine RPM or its multiple harmonics. This results into the resonance during machine normal operation and hence leads to frequent failures due to high vibration. The high vibration problem during installation and commissioning should not be solved using a trial-and-error approach. It is better to understand the dynamics of the as-installed machines.

Therefore it is highly recommended to do the additional tests (e.g., modal tests and/or Operational Deflection Shape analysis (ODS), etc.) on the machine during the installation and commissioning. The schematic of the Phase-1 vibration measurements is given in Figure 3.3. Modal tests and the ODS analysis are discussed in Chapters 8 and 9 with a few industrial examples. These tests

Vibration-Based Condition Monitoring and Fault Diagnosis

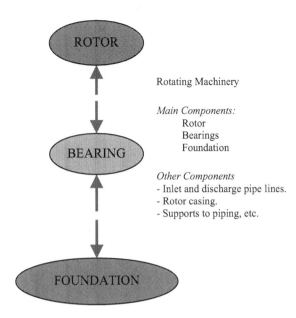

FIGURE 3.2 An abstract representation of a rotating machinery.

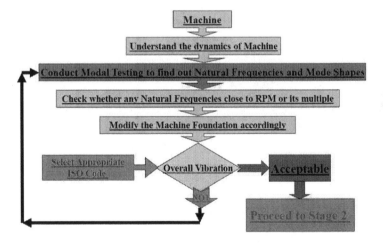

FIGURE 3.3 Stage 1: Vibration measurements during machine installation and commissioning.

can identify the likely root cause for the high vibration and possible reasons for frequent failure problems during the installation and commission stage itself. Hence the time span, t_1 for the Early Failure can be reduced which automatically enhances the *Useful life* time span, t_2 of the bath tub curve in Figure 3.1. Once the vibration problem is solved, it is better to proceed to the Stage-2 VCM as shown in Figure 3.3.

3.2.3 STAGE 2—MACHINE OPERATION

When it is confirmed that the vibration of machine is within the acceptable vibration limit during machine installation and commissioning then it is better to apply a regular VCM approach. This is Stage 2 of useful life of the machine as per the bath tub concept shown in Figure 3.1. It is expected that the machine should run safely and smoothly for many years but it does not means that the machine will not have trouble or failures. Few defects may randomly develop during machine regular operation so their early identification is important to rectify the identified defect during the planned shutdown. The steps involved in the VCM are shown in Figure 3.4. Each step is explained here.

1. *Machine*: Following details of machine must be handy before planning or carrying out vibration measurements.
   - Power rating so that appropriate ISO code can be selected for the vibration severity limit. ISO 10816 and 20816 codes may be useful for this purpose.
   - Number of bearings and their types. If they are anti-friction bearings then their full details must be known to estimate the defect-related bearing characteristics frequencies (refer Chapter 7). The clearance, type, and other details in case of fluid bearings.

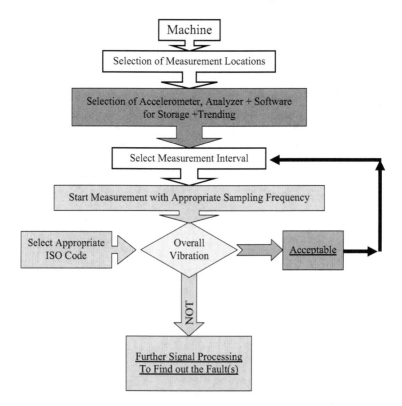

FIGURE 3.4 Stage 2: Abstract of a VCM procedure during the machine useful life.

Vibration-Based Condition Monitoring and Fault Diagnosis

- Number of impeller blades and number of stages.
- Details of gearbox if any.
- Machine RPM.
- Motor specifications if the machine is driven by the motor.

2. *Measurement locations and their directions*: ISO codes recommend the measurement on bearing housing/pedestal in three orthogonal directions. It is preferred location for the measurement as the measured vibration responses are expected to have responses from the rotor, bearing, gearbox, etc. The measurements in three orthogonal directions are also likely to map the dynamic behavior of any machine, which helps to detect any defects in machine at early stage.

For horizontal machines, the measurement should be taken in the axial, vertical and lateral directions. It is typically shown in Figure 3.5 through a photograph of a ball beating with housing. For vertical machines, the measurements should be in the axial and two-lateral directions as shown schematically in Figure 3.6.

FIGURE 3.5 Vibration measurements at bearing pedestal in vertical, horizontal, and axial directions for the horizontal rotor or machine.

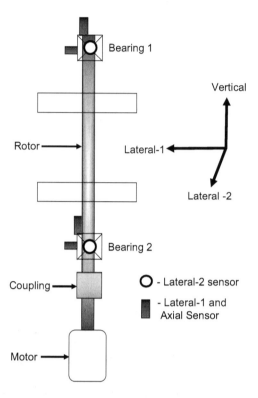

FIGURE 3.6 Vibration measurements at bearing pedestal in axial and 2-lateral (90 degree apart) directions for the vertical rotor or machine.

3. *Software and instrumentations*: Appropriate selection of instrumentation such as sensors, data collections and their analysis either through hardware or software codes are essential for the meaningful VCM. This is summarized in Chapter 11 using the required information from remaining chapters of the book.
4. *Measurement interval*: This is often found to be a topic of discussion within industries. It is difficult question to answer and hence it is difficult to decide whether the measurement is required continuously using online system or once per day or at the interval of few days. A few guidelines are discussed here that may help in deciding the measurement interval.

The P-F (Potential Failure) curve is well-known concept that represents the behavior of asset conditions from the "initiation of defect" to the "failure." A representative P-F curve for a machine is shown in Figure 3.7. It is difficult to have or predict the exact P-F curve for any machines or any components of a machine. However it is obvious from the P-F curve the VCM has the capability to identify the fault at much early stage and provides the lead time to failure ranging from a month and more. This simply means that the machine is not going to fail immediately even if a fault has already initiated. It is because it will go through a required number of vibration cycles before

Vibration-Based Condition Monitoring and Fault Diagnosis

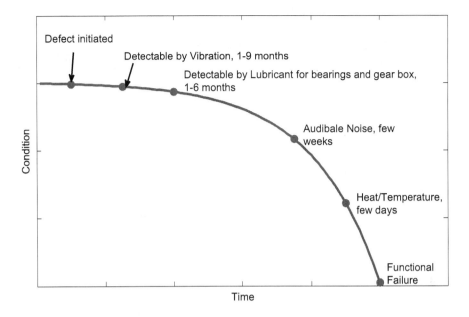

FIGURE 3.7 P-F curve.

leading to fatigue failure. Therefore a few days interval between measurements should be acceptable definitely for a portable VCM approach for any machines. If any machine is showing increasing trend in the vibration then it is better to do more frequent measurements to find reason for such behavior.

5. *Vibration severity limits*: It is essential step to define the allowable and alarm limits of the machine overall vibration for any VCM system. ISO code provides these limits for different machines. They recommend the overall vibration limits in the RMS velocity (V_{RMS}) of machine vibration. There are many ISO codes on this topic but ISO 10816 (1995) and ISO 20816 (2016) are likely to be useful for most of the conventional machines (such as rotating and reciprocating machines) used in practice. Table 3.1 is simplified to make it easy to understand the ISO code on vibration severity

TABLE 3.1
A Simplified Form of the ISO Table for Vibration Severity Limits for Rotating Machines

V_{RMS}, mm/s	Machine Classification	Vibration Limits
V_1	Categories based on power rating	Allowable
V_2		Satisfactory
V_3		Unsatisfactory
V_4		Unacceptable

limits for rotating machines. Table 3.1 gives four limits of vibration amplitude in V_{RMS}, mm/s —V_1, V_2, V_3, and V_4 as the *Allowable, Satisfactory, Unsatisfactory* and *Unacceptable limits* respectively. These vibration limits are divided into different classes or categories that depend on the power rating of machines in kW and MW, e.g., a pump driven by 10 kW motor is within the class of 15 kW machine (ISO 10816-1, 1995). It is user choice to use Allowable or Satisfactory Limit as the *Alarm Limit* for their machines. Chapter 7 explains why ISO codes have recommended the machine vibration velocity; however, the frequency range or band for the vibration measurement for any machines is not readily available in any ISO codes. The use of correct frequency band in the vibration measurement is very important that covers all fault frequencies (see details in Chapter 11) so that the measured RMS velocity fully reflects the machine conditions.

For example, the correct frequency range for a pump driven by a 2 kW motor is up to 6 kHz. The measured vibration velocity, V_{RMS} is 1 mm/s (above allowable limit of 0.71 mm/s) when measured up to 6 kHz. This means that the machine is operating above the allowable vibration limit and needs further investigation. But if the measurement is done only to 2 kHz, and the measured V_{RMS} is found to be only 0.5 mm/s, the machine is considered to be operating within the allowable vibration limit, and hence it is safe to operate. This is definitely not correct; therefore, it is always suggested to choose the frequency range for any machines carefully prior to the vibration measurements.

It is also important to note that there are many machines used in industries that may not be categorized by the ISO codes. It is the machine manufacturer's responsibility to provide the vibration severity limits for those machines or the users need to establish the allowable and alarm limits based on day-to-day experience with machines for the safe operation.

3.2.4 STAGE 3—AGED MACHINES

"Aged machine" seems to be a very subjective term. It is often observed that many identical machines within a plant or different plants within an organization continue to function properly even after many years of operation but some of them starts failing frequently e.g., frequent failure of bearings, gearbox, etc. after 15–20 years of operation. Hence it becomes difficult to define these frequently failing machines as the "aged machines" because other identical machines are still working in good conditions. This simply indicates that the some structural change may have happened over the period of operation which has resulted into the frequent and repeated failures. The regular VCM (Stage 2) is still good for these machines so that the defects can be predicted in time for remedial action before any catastrophic failure. In such situations, it is suggested to do a few tests similar to the Stage-I vibration measurements to identify the root cause of the frequent failures to solve the problems. This way the life of a machine can further be extended as shown in Figure 3.8. A couple of typical examples of frequently failing old machines are discussed in Chapters 8 and 9.

Vibration-Based Condition Monitoring and Fault Diagnosis

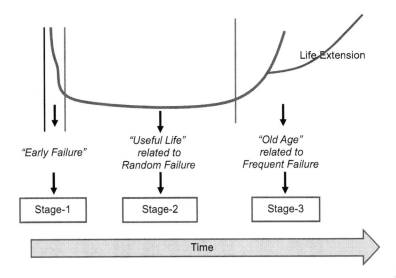

FIGURE 3.8 Bath tub Stage 3 modification—a possibility of the machine life extension.

3.3 SUMMARY

The chapter discussed the three different stages of machines and structures life like human beings—infant, young (useful life), and old age. The chapter also suggested three stages of vibration measurements during the machines life cycle to get the useful and reliable performance without impacting the plant safety and maintenance overheads.

REFERENCES

ISO 10816-1. Mechanical vibration: Measurement and evaluation of machine vibration Part 1: General guidelines, 1996.

ISO 20816-1. Mechanical vibration: Measurement and evaluation of machine vibration Part 1: General guidelines, 2016.

4 Vibration Instruments and Measurement Steps

4.1 INTRODUCTION

The equilibrium dynamic equation of a system shown in Figure 4.1 at time, t, is written as

$$F_i(t) + F_c(t) + F_s(t) = F(t) \tag{4.1}$$

$\Rightarrow$ Inertia Force + Damping Force + Stiffness Force = Applied Force

$$\Rightarrow \text{m } a(t) + \text{c } v(t) + \text{k } y(t) = F(t), \tag{4.2}$$

where $a(t)$, $v(t)$, and $y(t)$ are respectively the acceleration, velocity and displacement of the SDOF system at time, t. This simply means that any vibrating object generally consists of three types of vibration responses, namely; acceleration, velocity and displacement responses. All these parameters are inter-related, which implies that the measurement of one parameter may be used to generate any of the other parameters through integration or differentiation.

4.2 SENSORS AND THEIR MOUNTING APPROACH

It is very common to get analog electrical signal from any sensor. It is also a standard approach to get voltage signal from any sensors irrespective of their working principles so that the sensors output can directly be recorded onto the computer or data logger system through data acquisition (DAQ) device. A typical combination of a sensor with its power supply unit is shown in Figure 4.2.

Now-a-days there are many commercial sensors that have integrated electrical/electronic circuits within the sensors. These sensors don't need separate power supply units. In fact, many commercial DAQ systems are generally equipped with the option to supply required input power to the circuit within the sensors to carry out the measurements and data recording simultaneously.

There are many different types of sensors are available commercially, however the most commonly used sensors in the VCM are only discussed here.

4.2.1 DISPLACEMENT SENSOR

Proximity probe (or Eddy current probe) is commonly used sensor for measuring the relative displacement of a moving metal object relative to the sensor position. This sensor actually measures the gap between the object and sensor face. A typical

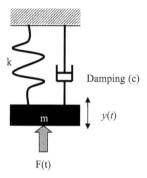

FIGURE 4.1 A spring-mass system.

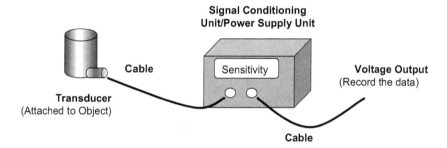

FIGURE 4.2 Vibration transducer (sensor) with a signal conditioning unit/power supply unit.

measurement arrangement is shown in Figure 4.3a and photograph of a couple of commercial proximity sensors is shown in Figure 4.3b. The gap between Objects 1 and 2 is nothing but the relative displacement of the vibration Object 1 with respect to the Object 2. A proximity probe assembly consists of a probe containing a coil as shown in Figures 4.3a. The coil is excited by an input current and voltage at a high frequency, typically 1 MHz or above. The high frequency of excitation is responsible for inducing eddy currents on the surface of the target object. The eddy current alters the induction of the probe coil, and this change can then translate into the voltage output, which is proportional to the distance between the probe and the vibrating object.

It is essential to know the sensor specification, typically minimum to maximum gap it can measure linearly as shown in Figure 4.3c so that the measurement can be planned carefully. This characteristic curve for the proximity sensor may change for different vibrating object materials.

This sensor is popularly used for measuring shaft relative displacement with respect to the bearing pedestal in rotating machinery. Another typical application is in the measurement of the reference signal (tacho signal) of a rotating shaft with a key phasor, with respect to the bearing pedestal (or non-rotating part) in rotating machinery. The tacho sensor is also discussed in Section 4.2.4.

Vibration Instruments and Measurement Steps

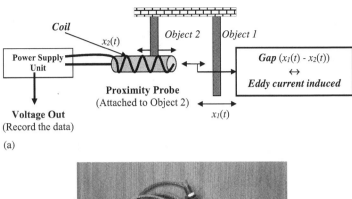

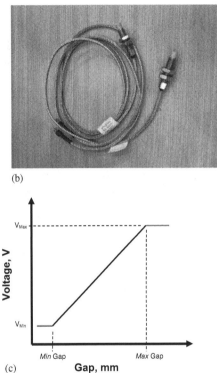

FIGURE 4.3 Proximity probes. (a) Measurement approach. (b) Photograph of two proximity probes. (c) Typical characteristics of proximity probe.

4.2.2 Velocity Sensor

Seismometer is a sensor to measure the vibration velocity and commonly used to measure the seismic events. It is generally bulky and hence not commonly used to measure machine vibration now-a-days. Hence this sensor is not discussed here but refers Sinha (2015) for further details.

Laser velocity sensor is another sensor for measuring vibration velocity. This is a laser-based sensor that works on the principle of the Doppler effect. It is a non-contact type sensor. The laser beam (pointer) from the sensor has to be focused on

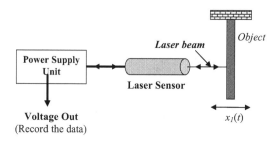

FIGURE 4.4 Vibration velocity measurement using a laser sensor.

the vibrating object. The change in wavelength of a reflecting laser beam, based on Doppler principle gives the velocity of the vibrating object. A simple schematic diagram of the measurement using a laser sensor is shown in Figure 4.4. Typical advantages of the laser vibrometer (sensor) are (i) measurement is possible from distance (no accessibility constraints), (ii) its non-contact nature ensures that the weight of the sensor is not added to that of the vibrating object, (iii) a single sensor can replace several contact type vibration sensors, by simply scanning the vibration of an entire area of the object, (iv) vibration measurement on small vibrating objects (up to micro-level) is also possible.

4.2.3 Acceleration Sensor

Several types of acceleration sensors (accelerometers) are available commercially. They are contact types (i.e., mounted directly on the vibrating object for acceleration measurement). The piezoelectric accelerometers are the most popular and most commonly used sensors in practice. A simple schematic construction of an accelerometer is shown in Figure 4.5. It consists of a mass (called seismic mass) mounted on a piezoelectric crystal and both are housed in a casing. Hence it is nothing but a spring-mass system where the piezoelectrical crystal is acting as a spring. The working of an accelerometer is schematically illustrated in Figure 4.6. The displacement, $y(t)$ of the vibrating object causes displacement, $y_e(t)$ of the accelerometer (seismic) mass so that the relative displacement, $y_{rel}(t) = y_e(t) - y(t)$ leads to a deformation in the piezoelectric crystal. This deformation will generate an equivalent amount of charge due to the piezoelectric property of the crystal, which is then converted to voltage signal, $v(t)$. This electrical output is known as the acceleration of the vibrating object. It is because this relative displacement is directly proportional to the acceleration of the vibrating object based on the vibration theory under certain

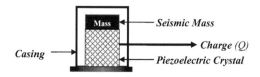

FIGURE 4.5 A simplified construction of an accelerometer.

Vibration Instruments and Measurement Steps

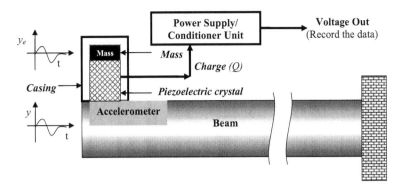

FIGURE 4.6 An accelerometer and its use.

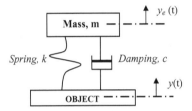

FIGURE 4.7 An SDOF System attached to a vibrating object.

conditions. Therefore the generated charge (or voltage) is directly proportional to the acceleration. The theoretical principle is explained here.

An accelerometer consists of an SDOF system. Figure 4.7 shows a simple configuration of a sensor mounted on a vibrating object. The equation of motion can be written as

$$m\ddot{y}_e(t) + c(\dot{y}_e(t) - \dot{y}(t)) + k(y_e(t) - y(t)) = 0 \qquad (4.3)$$

where $y(t)$ and $y_e(t)$ are the displacements of the vibrating object and the sensor mass (seismic mass), m respectively. Equation (4.3) can further be written as

$$m\ddot{y}_{rel}(t) + c\dot{y}_{rel}(t) + ky_{rel}(t) = -m\ddot{y}(t) \qquad (4.4)$$

where $y_{rel}(t) = y_e(t) - y(t)$ is the relative displacement of the sensor seismic mass with respect to the displacement of the vibrating object. The solution of Equation (4.4) is then written as

$$y_{rel}(t) = \frac{-\left(\ddot{my}(t)/k\right)}{\sqrt{(1-r^2)^2 + (2\zeta r)^2}} = \frac{-\left(\ddot{y}(t)/\omega_n^2\right)}{\sqrt{(1-r^2)^2 + (2\zeta r)^2}} \qquad (4.5)$$

50 Industrial Approaches in Vibration-Based Condition Monitoring

where $r = \frac{\omega}{\omega_n} = \frac{f}{f_n}$, f is the frequency of the vibration object and f_n is the natural frequency of the sensor. If the ratio of the frequency, $r = \frac{\omega}{\omega_n} = \frac{f}{f_n} <<< 1 \approx 0$ then Equation (4.5) can be written as

$$y_{rel}(t) = -\frac{\ddot{y}(t)}{\omega_n^2} \propto \text{Acceleration of vibrating object} \tag{4.6}$$

Hence when $r = \frac{\omega}{\omega_n} = \frac{f}{f_n} <<< 1 \approx 0$ then the relative motion, $y_{rel}(t)$ of sensor mass with the vibrating object is directly proportional the acceleration, $\ddot{y}(t)$ of the vibration object because ω_n^2 is the constant term for the accelerometer. Since the spring of accelerometer is made of a piezoelectric crystal as shown in Figure 4.5 so the relative displacement, $y_{rel}(t)$ in the piezoelectric crystal is going to generate an equivalent amount of the charge, $Q(t)$. Therefore

$$\text{Relative displacement} \quad y_{rel}(t) \propto Q(t) \propto \text{Acceleration of Vibrating Object} \tag{4.7}$$

where the charge, $Q(t)$ is the measure of the acceleration which can then be converted into the voltage.

However, it is also important to understand the impact of the condition of $r = \frac{\omega}{\omega_n} = \frac{f}{f_n} <<< 1 \approx 0$ on the acceleration measurement. Equation (4.5) can be written in the non-dimensional form as

$$\frac{\omega_n^2 y_{rel}(t)}{\ddot{y}(t)} = -\frac{1}{\sqrt{\left(1-r^2\right)^2 + \left(2\zeta r\right)^2}} \tag{4.8}$$

where $y_{rel}(t)$ is the output (OP) and the vibrating object acceleration, $\ddot{y}(t)$ is the input (IP) of the accelerometer. The damping ratio and the frequency ratio (r) cannot be ignored in reality.

Figure 4.8 shows the characteristic curve of an accelerometer in terms of the non-dimensional form of the transfer function (OP/IP) for different values of damping ratios. It is obvious from the (OP/IP) plot in Figure 4.8 is nearly equal to 1 for a very small value of the frequency ratio, r. This means that the working frequency range of an accelerometer may be a small fraction of its natural frequency, f_n. It is generally 20% of f_n when the damping ratio is close to 0.7. So if the natural frequency of an accelerometer is 25 kHz then it can measured the vibration correctly up to $0.20 f_n = 5$ kHz only. It is always important to check the accelerometer specifications before selection of an accelerometer for your machine vibration measurement (refer Chapter 11). Few typical specifications are listed here.

- Sensitivity: pC/g, pC/(m/s²), mV/g, mV/(m/s²).
- Frequency range of measurement
- Resonance or natural frequency
- Maximum acceleration measurement limit

Vibration Instruments and Measurement Steps

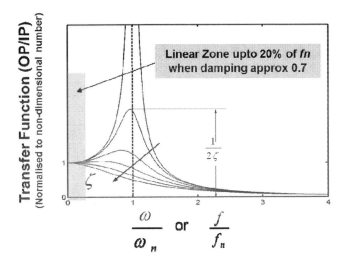

FIGURE 4.8 Characteristics curve of an accelerometer in non-dimensional form of the transfer function (OP/IP).

- Amplitude linearity—This is the ratio between the measured acceleration to the actual acceleration.
- Phase distortion—There should not be any phase error between the measured acceleration and the actual acceleration of the vibrating object.

Specifications of a typical accelerometer are also listed in Table 4.1 for clarity.

Accelerometer mounting: It is known from theory that up to 20% of the accelerometer natural frequency, f_n is only useful for accurate measurement of the acceleration of any vibrating object. However, when an accelerometer is mounted on an object, the mounting resonance frequency, f_m of the accelerometer may vary. This generally depends on the types of mounting used. There are several mounting techniques (wax, adhesive, magnet or stud), with each type possessing its own stiffness. The mounting stiffness may get added to the accelerometer stiffness in series. It may have some other influences but they may not be significant and hence ignored here, so the effective stiffness, k_{eff} for the accelerometer can be given by

$$\frac{1}{k_{eff}} = \frac{1}{k} + \frac{1}{k_m} \tag{4.9}$$

where k_m is the stiffness of the mounting type which may be equal to, or greater than or less than the stiffness, k of the accelerometer. The effective stiffness, k_{eff} may be less than or equal to k and therefore, the mounting natural frequency, $f_m = \frac{1}{2\pi}\sqrt{\frac{k_{eff}}{m}} \leq f_n$. This means that the working frequency range of the accelerometer may be up to $0.2 f_m$ which may be less than or equal to $0.2 f_n$. This is illustrated in Figure 4.9. Figure 4.9 also compares the possible trend in the changes in the natural frequency of an accelerometer due to different mounting arrangements. This simply

TABLE 4.1
Typical List of Accelerometer Specifications

Parameters	Values	Remarks
Sensitivity	100 mV/g	Generates 100 mV for 1 g (or 9.81 m/s) vibration acceleration input.
Measurement range	±50 g	Limits of maximum and minimum level of vibration acceleration measurement.
Resonance frequency	≥50 kHz	This is the natural frequency of the accelerometer. For exact value, it is better to ask for test/calibration chart.
Frequency range (±5%)	0.5 Hz–10 kHz	10 kHz is acceptable as per the theory of $0.20 f_n$ but it can have amplitude error up to ± 5%.
Resolution	0.00015 g	This is the least count for the accelerometer i.e., the minimum fluctuation in vibration can be measured accurately.
Overload limit (shock)	±5000 g (peak)	Accelerometer may not get damaged up to this load.
Operating temperature range	−65° to +200°F	Accelerometers with different temperature range are also available commercially to meet industrial requirements.
Excitation voltage	18–30 V (DC)	Either a separate power unit for this accelerometer is need or the DAQ device should have this option.
Constant excitation current	2–20 mA	
Size	Select as per the requirements.	
Weight		

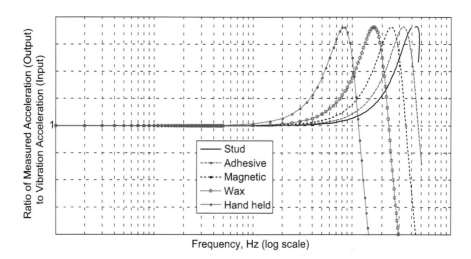

FIGURE 4.9 Impact of different mountings on accelerometer behavior.

Vibration Instruments and Measurement Steps

indicates that even if an accelerometer is good for the job but it may or may not provide accurate measurements. The reduction in the linear frequency range of the accelerometer is main reason for the possible error in the measurements in many applications. Hence the use of the appropriate mounting depending upon the applications is important consideration in vibration measurements.

4.2.4 Tacho Sensor

In a rotating machine, a reference signal is required to measure rotating speed and the phase of rotating shaft and its vibration with respect to the reference signal. The reference signal measured by a sensor is call tacho sensor. The proximity probe and laser pointer sensor are the most commonly used tacho sensors.

Two measurement schemes are shown in Figure 4.10. Proximity probes are used to measure the gap between the shaft outer surfaces to the probe head in both cases. The shaft vibration is ignored during its rotation in both cases. In Case (i), the gap measured by the proximity probe will remain constant with time during the shaft rotation as shown in Figure 4.11a and hence it is difficult to measure the shaft speed. In Case (ii), however, there will be step change in the gap due to the presence of keyway in the shaft during each shaft rotation. The tacho signal for Case (ii) is shown in Figure 4.11b; therefore, it is possible to measure the time (T) for 1 complete shaft rotation. Once the time (T) in second is known, then the shaft speed can be estimated as 1/T rotation per second (RPS) or 60/T rotation per minute (RPM).

Similarly the laser pointer sensor can also be used for this purpose. It may be better option compared to the proximity probe. No keyway is required in the shaft. Instead a small reflective tape can be used on the shaft surface as a target reference, which can then be tracked from a distance by a laser pointer sensor. Typical measurement arrangement on a rotating rig is shown in Figure 4.12. A small reflective tape functions as the keyway in the shaft. The voltage output from the laser sensor is going to be different when the laser beam pointer is passing through

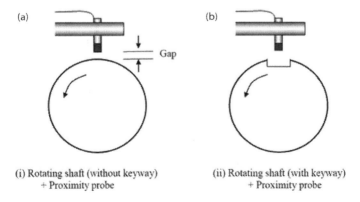

FIGURE 4.10 Concept of tacho sensor (a) without keyway (b) with keyway on the shaft.

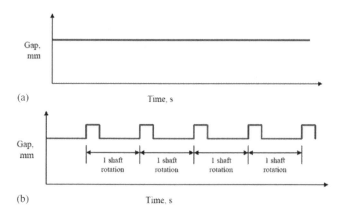

FIGURE 4.11 Measurement by the tacho sensor (a) without keyway (b) with keyway on the shaft.

FIGURE 4.12 Photographs of laser tacho sensor and its use in rotating machine.

the reflective tape. This is illustrated in Figure 4.13. The shaft is rotating clockwise. A "dot" is marked on the shaft (side view) at the edge of the reflective tape. The measured tacho signal during the shaft rotation is also shown in Figure 4.13. The sudden increase in the tacho signal voltage in the signal indicates one edge of the reflector and then a drop in voltage another edge of the reflection. If the first rise in the voltage is assume to be starting point of the shaft rotation (which is also marked by a dot) then the time between two dots on the tacho signal indicates one complete rotation (360°) of the shaft. If the dot on the shaft is assumed to be "zero" degree for the shaft with respect to the tacho reference point then it is easy to map the angle of shaft during machine rotation as shown in Figure 4.13. It is important to keep the positions of both reflector and tacho sensor exactly same all the time for any comparison purpose. The use of this tacho signal in the rotor balancing is discussed in Chapter 7.

Vibration Instruments and Measurement Steps

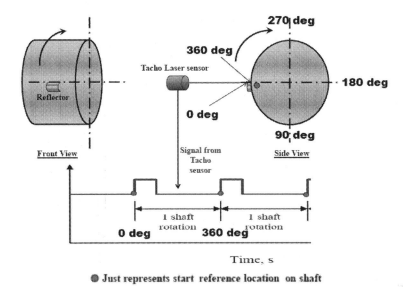

FIGURE 4.13 Speed and phase angle of the shaft with respect to the tacho sensor.

4.3 VIBRATION MEASUREMENT

4.3.1 A Typical Measurement Setup

A typical example of vibration measurement setup on a steam turbo-generator (TG) set is shown in Figure 4.14. The schematic of the TG set shown in Figure 4.14 (Sinha, 2002) consists of a high-pressure (HP) turbine, a low-pressure (LP) turbine and an electric generator. In some high-capacity power plants the TG sets also have a few intermediate pressure (IP) turbines and more number of LP turbines that are not shown in Figure 4.14. The shafts of the individual systems are joined together by means of couplings, and the complete shaft is known as the rotor of the TG set. The shafts of each turbine are designed to have a number of rows of turbine blades along the shaft length. Each rotor of an individual system is normally supported by its own fluid journal bearings, which are supported on foundation structures which are often flexible.

Experience shows that faults develop in rotating machines during normal operation, for example bends, cracks or mass unbalance in the shaft (due to scale deposits and/or erosion in blades of the turbines). If a fault develops and remains undetected for some time, then, at best, the problem will not be too serious and can be remedied quickly and cheaply. However, at worst, it may result in expensive damage and downtime, injury, or even loss of life. Such a situation warrants the use of a reliable condition monitoring technique to reduce the down time of the plant.

Condition monitoring requires the continuous observation of various parameters like vibration responses at different locations of the TG sets and many other process parameters, for example, pressure and temperature of fluid bearing, steam,

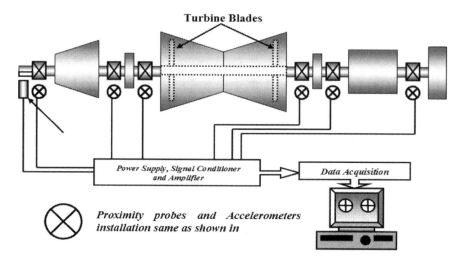

FIGURE 4.14 Schematic of vibration response measurement setup on a turbo-generator (TG) set.

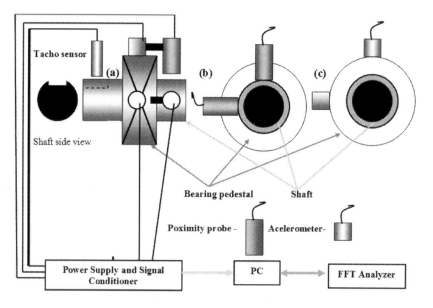

FIGURE 4.15 Vibration sensors typically at bearing of the TG set shown in Figure 4.14 with a tacho sensor.

condenser, during the machines normal operation. Hence only vibration-based condition monitoring (VCM) is discussed here.

Figure 4.14 also presents the schematic the vibration measurements (with details of sensors at each bearing in Figure 4.15) together with sensor power supply/conditioner units, data logger and data acquisition (DAQ) system for collecting the measured data into the computer. It is obvious from Figures 4.14 and 4.15 that there are number of steps involved in measurements which are explained here.

4.3.2 Steps Involved in the Data Collection

It is vital to understand and/or have knowledge of the following prior to setting up a measurement/monitoring system (from a vibration engineer's point of view).

1. Specifications (specs.) of object (machine, structure, etc.) on which vibration measurements need to be carried out
2. Knowledge of sensors' working principle (not the exact electrical/electronic circuits)
3. Knowledge of sensors' specifications, so as to aid appropriate selection
4. Knowledge of how to collect the data into a computer
5. Signal processing to interpret the measured data
6. Selection of data display to monitor the machine condition

A typical abstract of the measurement procedure step-by-step is shown in Figure 4.16 and explained systematically for the purpose of clarity.

1. A number of different sensors depending on the requirements. P indicates the measuring parameter from a sensor (e.g., P will be displacement in mm for the displacement sensor). "S_o" indicates the sensor output possibly in voltage, depending upon the sensor sensitivity, S.
2. Conditioner/power supply unit for each sensor, which supplies power to the sensor and provides an analog voltage output. Currently the power supply circuit is integrated within the sensor itself for many commercially sensors.

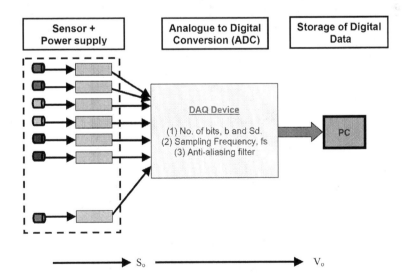

FIGURE 4.16 Typical abstract form of any measurement procedure.

3. Analog voltage output from each sensor (or sensor and power supply unit) is then connected to the data acquisition (DAQ) device, which converts the analogue signals from different sensors into a digital form.
4. Select appropriate DAQ device with required number of bits (b) for analog-to-digital conversion (ADC), DAQ sensitivity, S_d, sampling frequency, f_s, number of input channels, and the anti-aliasing filter. Photographs of few DAQ devices are shown in Figure 4.17.
5. V_o indicates the recorded voltage digital data from each sensor into the PC.

Once the data are recorded, signal processing can then be performed to extract the appropriate information. The signal processing is discussed in Chapter 5.

FIGURE 4.17 A few multi-input channels DAQ devices.

Vibration Instruments and Measurement Steps

4.3.3 INSTRUMENT CALIBRATION AND SPECIFICATIONS

Following considerations are vital for the selection of the appropriate sensor(s) and instrumentation.

- *What kind of output do you require?*—This will depend on the parameter (e.g., temperature, pressure, etc.) to measure.
- *What level of accuracy is required?*—The measurement error should be within 5% or 1% or less. This needs to be judged based on applications and requirements.
- *What level of resolution required?*—A step resolution in the measuring parameter that the sensor can accurately measure.
- *Is the sensor sensitive to environmental conditions?*—It is also important to know whether the sensor is sensitive to environment conditions such as variations in temperature.
- *What is the range of measurement needed?*—The upper and the lower limits for the measurement to meet the requirements.
- *What happens to the sensor if the measured parameter exceeds the limit?*— It is required to know what happens to the sensor if the measurement exceeds the limit. Will this cause damage to the sensor, or does the sensor have an inbuilt protection capability?
- *What is the mounting arrangement?*—The size, weight and mounting arrangements are important considerations, so that this should not affect the normal setup and/or the required function of the system on which experiments are to be conducted.
- *Is portable type or permanent mounting best?*—Sometimes, this becomes an essential consideration.

The consideration of the above aspects is only possible if the following technical specifications are carefully looked into.

- *Calibration chart*: It is a standard practice that newly manufactured sensors have to pass through a series of tests to check their performance and quality. One of the important tests is calibration. During this test, the performance of the sensor is tested against a known input or by comparing its output to the output from a well-accepted reference sensor. Such a calibration chart is often considered a valuable document for any sensor.
- *Sensitivity (electrical parameter/measuring parameter)*: Currently, all of kinds of sensors are associated with an electrical/electronic circuit which provides electrical output, generally in voltage. Hence, the ratio of the electrical output to the parameter measured by the sensor is generally known as the sensitivity of the sensor. For example, sensitivity is 10 mV/mm for a displacement sensor means that an output of 10 mV represents a displacement of 1 mm.
- *Range of measurement (lower and upper)*: The lower and upper bound of the parameter to be accurately measured by a sensor is called the range

of measurement. The difference between the measurable upper and lower limits is called full scale (FS) reading of the instrument.

- *Linearity or accuracy*: The accuracy or linearity of an instrument is the measure of how close the output reading of the instrument is to the actual value. In practice, it is common to quote the inaccuracy figure rather than the accuracy figure. Inaccuracy is the extent to which the reading might be wrong. It is often quoted as a percentage of the FS. For example, a pressure gauge of range 0–10 bar has a quoted inaccuracy of $\pm 1\%$ FS, which means that the maximum error is expected to be 0.1 bar.
- *Repeatability/reproducibility*: Repeatability describes the closeness of the output reading when the same input is applied repetitively over a short period of time, while keeping all measurement conditions (instrument, observer, location, usage, etc.) same. Reproducibility, on the other hand, describes the closeness of the output readings for the same input when there are changes in the method of the measurement, observer, measuring instrument, measurement locations, conditions of use and time of measurement. The degree of repeatability or reproducibility in measurements from an instrument is an alternative way of expressing its precision.
- *Stability*: The output of a sensor should remain constant when a constant input is applied over long periods of time, compared with the time of taking a reading, under fixed conditions of use.
- *Resolution*: This is smallest value or smallest change of the measuring parameter that the instrument can measure with accuracy.
- *How sensitive is the sensor under different environmental conditions?*: Calibration is usually conducted under a fixed condition, e.g., constant temperature, humidity, etc. Hence it is good to know the effect of different environment conditions on the accuracy of the measurements.
- *Over-range protection*: It is likely that selection of a sensor/instrument may not be accurate and in case the measuring parameter is more than the limit of the sensor, the sensor may be damaged and cannot be used further. It is good to have a range protection system within the sensor. Sometimes, the measured parameter could exceed the limits of the sensor, due to inaccurate sensor/instrument selection. This can cause a permanent damage to the sensor/instrument. It is therefore desired to have a protection system within the sensor.
- *Shock withstanding capacity*: Sensors should have in-built shock protection capabilities, so as to prevent damage/faults when accidentally dropped from their mounting locations.
- *Signal conditioner and/or power supply unit*: Each sensor requires some kind of power supply to activate its circuit and receive the voltage output from the sensor. It is good to understand whether this unit is required separately or already integrated with the sensor and compatible with DAQ device.
- *Size, weight, mounting guide, etc.*: Size and weight of the sensors are also important considerations when performing experiments. It is good to refer to the mounting guidelines from the product catalogue, the manufacturer's notes and accessibility to the measurement location.

Vibration Instruments and Measurement Steps

4.3.4 Concept of Sampling Frequency

Figure 4.18 is a typical analog voltage signal from a sensor. Generally, all the data in any continuous analog signal are difficult to collect into the PC. Hence a definite rate of data collection needs to be defined, e.g., 20 samples/s. This means that 20 data samples from the signal will be collected in a second. This is known as the data sampling rate or sampling frequency, f_s. This means that each data is collected at equal intervals of $1/20 = 0.05$ s. The time interval or time step (dt) between two constitutive samples of the data collection becomes equal to 1/fs. If the data length is 5 s, then a total of 101 samples (1 sample at 0 s + 20 samples × 5 s) will be collected into the computer. This process is a part of the analog-to-digital conversion (ADC) process.

Let's consider an original analog voltage signal from a sensor shown in Figure 4.18. Now it is considered to be collected at the sampling frequency, $f_s = 20$ samples/s (or Hz). The signal data will then be sampled in the digital form in the following way as listed in Table 4.2 for the original signal in Figure 4.18.

Figure 4.19a and b show the re-constructed signals after data collection into the PC at the sampling rates, $f_s = 20$ Hz and $f_s = 40$ Hz respectively from the original analog signal in Figure 4.18. It is obvious that the reconstructed signals are significantly different from the original signal. It can only be improved by increasing the data sampling rate, so an appropriate selection of the sampling frequency, f_s is important during the process of data collection.

It is also important to note that the recorded voltage at each time step may or may not be exactly same as the voltage of the original signal. This depends on the number of bits of the DAQ device. This is explained in Section 4.3.7.

4.3.5 Aliasing Affect and Anti-aliasing Filter

To demonstrate an aliasing affect, an analog sinusoidal signal at a frequency, $f = 500$ Hz from a sensor is considered here. The signal is shown in Figure 4.20. Now this signal is recorded into a computer through a DAQ device but with different sampling frequency, f_s. The aim is to observe the impact of the sampling frequency on collection of data from the sinusoidal signal in Figure 4.20.

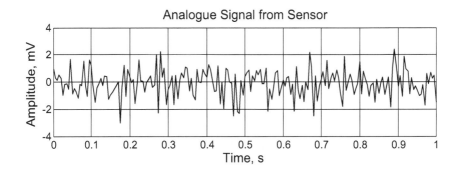

FIGURE 4.18 An original analog signal.

TABLE 4.2
Sampled Data for the Signal in Figure 4.18 at $f_s = 20$ Hz

Data	Time, s		Signal Amplitude, mV
0	0	0	0.9492
1	dt	0.0500	−0.8757
2	2 dt	0.1000	1.2347
3	3 dt	0.1500	−0.9480
4	4 dt	0.2000	0.3503
5	5 dt	0.2500	0.2323
6	6 dt	0.3000	−0.5900
7	7 dt	0.3500	1.0061
8	8 dt	0.4000	0.3502
9	9 dt	0.4500	1.0205
10	10 dt	0.5000	0.4115
11	11 dt	0.5500	−0.1472
12	12 dt	0.6000	0.1370
13	13 dt	0.6500	−0.2539
14	14 dt	0.7000	0.1644
15	15 dt	0.7500	−0.7841
16	16 dt	0.8000	1.4790
17	17 dt	0.8500	−0.8223
18	18 dt	0.9000	−0.3158
19	19 dt	0.9500	−0.5700
20	20 dt	1.0000	−1.4831

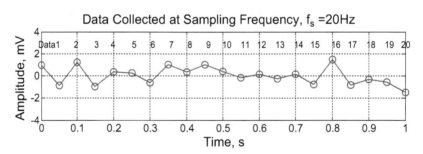

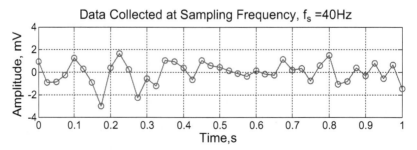

FIGURE 4.19 Collected (sampled) signals using different sampling frequencies.

Vibration Instruments and Measurement Steps

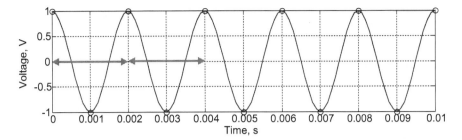

FIGURE 4.20 A typical sine wave of frequency 500 Hz.

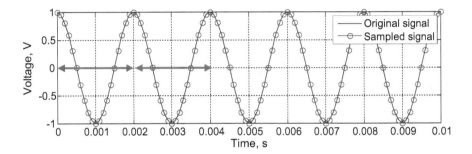

FIGURE 4.21 Original and sampled signals when $f_s = 10$ kHz.

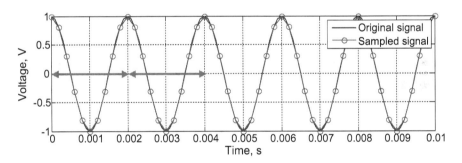

FIGURE 4.22 Original and sampled signals when $f_s = 5$ kHz.

Case (i): The sampling frequency, $f_s = 10$ kHz, dt = 0.1 ms. The collected data (marked as a solid line with circles) shown in Figure 4.21 is same as the original data in Figure 4.20. Hence there is no error in the collected/recorded data.

Case (ii): The sampling frequency, $f_s = 5$ kHz, dt = 0.2 ms. The collected data (marked as solid line with circles) is compared with the origin data (solid line) in Figure 4.22. The collected/recorded signal is nearly same as the original signal.

Case (iii): The sampling frequency, $f_s = 2$ kHz, dt = 0.5 ms. The collected data (marked as solid line with circles) is compared with the origin data (solid line) in Figure 4.23. The time period of 0.002 s (marked by double arrows) of the recorded signal is same as the original signal. The maximum and minimum voltages of the recorded signal is also same as the original signal but the number of samples per second were not enough to collect a signal that is a replica of the original signal.

Case (iv): The sampling frequency, $f_s = 1$ kHz, dt = 1 ms. The observation shown in Figure 4.24 is not much different from case (iii).

Case (v): The sampling frequency, $f_s = 750$ Hz, dt = 0.00133 s. The recorded signal (marked as solid line with circles) is compared with the origin signal (solid line) in Figure 4.25. The time period of the recorded signal is 0.004 s

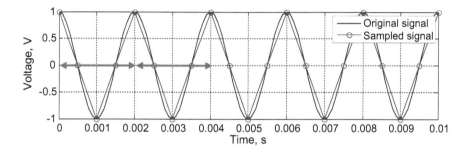

FIGURE 4.23 Original and sampled signals when $f_s = 2$ kHz.

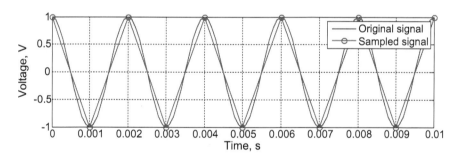

FIGURE 4.24 Original and sampled signals when $f_s = 1$ kHz.

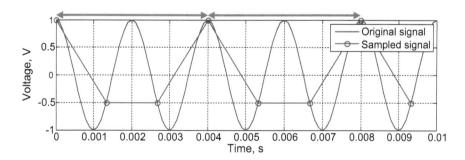

FIGURE 4.25 Original and sampled signals when $f_s = 750$ Hz.

Vibration Instruments and Measurement Steps

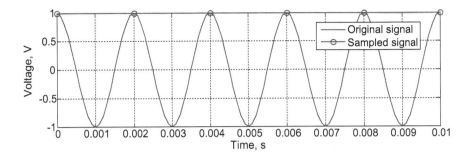

FIGURE 4.26 Original and sampled signals when $f_s = 500$ Hz.

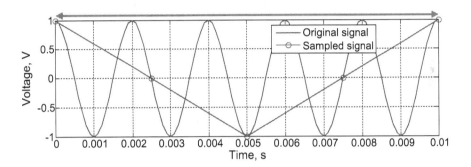

FIGURE 4.27 Original and sampled signals when $f_s = 400$ Hz.

which is twice that of the original signal, 0.002 s. Significant error in amplitude is also observed in the recorded signal.

Case (vi): The sampling frequency, $f_s = 500$ Hz, dt = 0.002 s. The recorded signal (marked as solid line with circles) is just a straight line compared to the origin sinusoidal signal (solid line) in Figure 4.26. This means that the time period for the recorded signal is "infinity." This is a typical case of synchronization of the signal frequency with the sampling rate, causing the frequency of the recorded signal to be "zero."

Case (vii): The sampling frequency, $f_s = 400$ Hz, dt = 0.0025 s. The recorded signal (marked as a solid line with circles) is compared with the origin signal (solid line) in Figure 4.27. The time period of the recorded signal is 0.01 s, which is much higher than the original signal, 0.002 s. Significant error in amplitude is also observed for the recorded signal.

4.3.5.1 Observations
- Original signal frequency, $f = 500$ Hz.
- Sampling frequency (f_s) much higher than $2f$ indicates no error.
- Sampling frequency (f_s) equals to $2f$ is likely to predict the frequency correctly but possible error in the amplitude.

- Sampling frequency less than $2f$ indicates that a "high frequency signal" (i.e., 500 Hz here) may appear as a "low frequency signal" with significant amplitude error—This is called aliasing during the A to D conversion (ADC) process during the data collection.

Therefore, the DAQ device must be equipped with the anti-aliasing filter. This filter is discussed in Section 4.3.6.

4.3.6 Concept of Nyquist Frequency, f_q and the Useful Upper Frequency Limit, f_u

It is clear from the discussion in Section 4.3.5 that the signal will be collected correctly only up to $f_s/2$ if the sampling frequency is f_s samples per second (Hz) during the ADC process. Hence the frequency $f_s/2$ is known as Nyquist frequency, f_q.

Most of the time the data collection is done blindly with a chosen sampling frequency, f_s. Therefore any frequency or frequencies above the Nyquist frequency, f_q in the signal will be sampled as the lower frequencies in the collected signal. This is definitely error in the collected signal and known as the aliasing effect. It is important to filter out all frequencies above the Nyquist frequency, f_q in the signal during the data collection process itself to ensure that the collected data contains correct information (without aliasing affect) up to the Nyquist frequency, f_q. Therefore the DAQ device must be equipped with a low pass filter, which can filter out all frequencies above the Nyquist frequency, f_q during the data collection. This low pass filter is known as the anti-aliasing filter.

Ideally, a sharp step-type digital anti-aliasing filter is needed at the Nyquist frequency, f_q as shown in Figures 4.28 and 4.29; however, it seems it is difficult to achieve in practice. Hence a standardized practice is to use a filter which gradually starts filtering out the frequencies from the frequency, $f_s/2.56$ to the Nyquist frequency, f_q i.e., reducing the amplitude of the signal to be collected from the frequency, $f_s/2.56$ to zero amplitude at the Nyquist frequency, f_q to eliminate aliasing affect in the collected signal. This filter is also shown in Figures 4.28 and 4.29. All the commercial

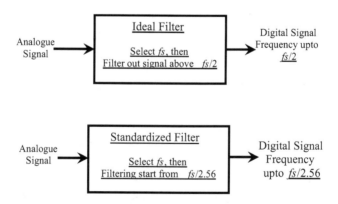

FIGURE 4.28 Anti-aliasing filters, ideal and standardized filters.

Vibration Instruments and Measurement Steps

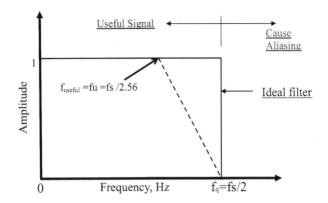

FIGURE 4.29 A simple representation of a standardized anti-aliasing filter used in practice.

instruments used this standardized gradual anti-aliasing filter from the frequency, $f_s/2.56$ to f_q if they are using anti-aliasing filter. Note that there are many commercial DAQ devices that are not equipped with anti-aliasing filter.

There will no change in the amplitude of the collected signal amplitude up to the frequency, $f_s/2.56$ after the ADC process with anti-aliasing filter from the original analog signal from a sensor. Therefore it is good to analyze that data only up to the frequency, $f_s/2.56$. This frequency is named here as the useful upper frequency limit, f_u. This means that if the frequency range up to 10 kHz is required for a machine vibration monitoring, then $f_u° = 10\,\text{kHz}, f_s = 2.56 f_u = 25.6$ kHz and $f_q = f_s/2 = 12.8$ kHz.

4.3.7 Analog-to-Digital Conversion (ADC)

Once the sampling frequency is selected, then an appropriate DAQ card is required. The ideal DAQ card should be capable of supporting the required number of input channels and the sampling frequency together with anti-aliasing filter per input channel for the simultaneous collection of data from all sensors.

DAQ is basically an electronic device where the analog-to-digital conversion (ADC) is generally done using a binary representation for encoding the signal. This means that the smallest value of a signal using a 12-bit binary representation would be represented as:

000000000000 = 0 and the largest value: 111111111111 = 12

Let's assume that 0 represents "0 V" and 12 represents "10 V," then the sensitivity of this device will be written as $S_d = (10-0)\text{V}/12 = 0.8333$ V/binary bin, where (10−0)V is the full scale input voltage (FSIV) and the total number of binary bins, BT = 12. The DAQ device works on the principle of converting the input voltage into the number of the nearest binary bins (B), depending on the DAQ device sensitivity, S_d. The bit (b) of the DAQ device/card indicates the total number of binary bins (BT) for the card, which is calculated as $BT = 2^b$. This means that $BT = 2^4 = 16$ binary bins for 4-bit DAQ device.

The sensitivity of the DAQ device is defined as the ratio of the full-scale input voltage (*FSIV*) to the total number of bins (*BT*). Mathematically it is expressed as

$$S_d = \frac{FSIV}{BT} \tag{4.10}$$

Consider a 16-bit DAQ card with two input channels. The input voltage level for each channel is assumed to be ±5 V. Hence the sensitivity (S_d) of the DAQ card is $\frac{FSIV}{2^b} = \frac{10}{2^{16}} = 0.15258$ mV/bin. Table 4.3 gives a summary of how the analog voltage signal from a sensor is digitize into number of bins and then to the voltage during the process of ADC at any selected sampling frequency.

To further aid the understanding, the example of an analog signal shown in Figure 4.18 is considered again. Figure 4.19 shows the sampled signal without any amplitude at the sampling frequency, $f_s = 20$ Hz is typically chosen for this demonstration. The accurate amplitude vales of the sample data are listed in Table 4.2 for the reference before converting into the binary bins and then to voltage during the ADC process.

TABLE 4.3

Rule for Analog-to-Digital Conversion Using 16-bit DAQ Device

Actual Signal Voltage Range	Digital Conversion (No. of Bins, B)	Re-construction of Signal from Bins (B × S_d)
−5 V to less than −(5−1.5259e−4)V	−32768	−5.00 V
−(5−1.5259e−4)V to less than −(5−2 × 1.5259e−4)V	−32767	−(5−1.5259e−4)V
−(5−2 × 1.5259e−4)V to less than −(5−3 × 1.5259e−4)V	−32766	−(5−2 × 1.5259e−4)V
⋮	⋮	⋮
−2 × 1.5259e−4 V to less than −1.5259e−4 V	−2	−2 × 1.5259e−4 V
−1.5259e−4 V to less than 0	−1	−1.5259e−4 V
0	0	0.00 V
Greater than 0 to +1.5259e−4 V	+1	+1.5259e−4 V
Greater than +1.5259e−4 V to +2 × 1.5259e−4 V	+2	+2 × 1.5259e−4 V
⋮	⋮	⋮
Greater than +(5−3 × 1.5259e−4)V to +(5−2 × 1.5259e−4)V	+32766	+(5−2 × 1.5259e−4)V
Greater than +(5−2 × 1.5259e−4)V to +(5−1.5259e−4)V	+32767	+(5−1.5259e−4)V
Greater than +(5−1.5259e−4)V to +5 V	+32768	+5.00 V

Vibration Instruments and Measurement Steps

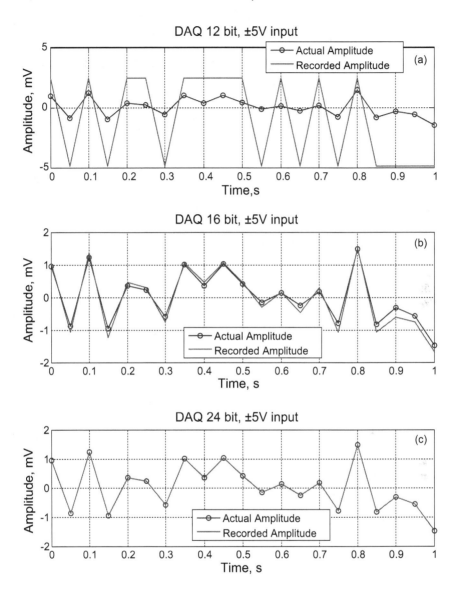

FIGURE 4.30 Recorded signal amplitudes with different DAQ devices (a) 12 bit, (b) 16 bit and (c) 24 bit.

Now three cases are considered for the ADC at the sampling frequency, $f_s = 20$ Hz. These cases are (a) 12-bit DAQ device, (b) 16-bit DAQ and (c) 24-bit DAQ and all having ±5 V input voltage for the analog input signals. The collected signals are shown in Figure 4.30 and their values are listed in Table 4.4 for easy comparison with the values in Table 4.2. It is clear from these three cases the accuracy in the collected signals is much better in case of 24-bit DAQ device. The amplitudes of data at all time steps (Table 4.4) are almost same as the actual values listed in Table 4.2 in case of 24-bit DAQ device.

70 Industrial Approaches in Vibration-Based Condition Monitoring

TABLE 4.4

Collected Data for the Signal in Figure 4.18 Using Different DAQ Devices

DAQ Properties	12-bit; $S_d = 2.441$ mV/bin		16-bit; $S_d = 0.1526$ mV/bin		24-bit; $S_d = 0.596\,\mu$V/bin	
Time, s	B	$V_0 = B \times S_d$	B	$V_0 = B \times S_d$	B	$V_0 = B \times S_d$
0	1	2.4414	7	1.0681	1593	0.9495
0.0500	−2	−4.8828	−7	−1.0681	−1471	−0.8768
0.1000	1	2.4414	9	1.3733	2072	1.2350
0.1500	−2	−4.8828	−8	−1.2207	−1592	−0.9489
0.2000	1	2.4414	3	0.4578	588	0.3505
0.2500	1	2.4414	2	0.3052	390	0.2325
0.3000	−2	−4.8828	−5	−0.7629	−991	−0.5907
0.3500	1	2.4414	7	1.0681	1688	1.0061
0.4000	1	2.4414	3	0.4578	588	0.3505
0.4500	1	2.4414	7	1.0681	1713	1.0210
0.5000	1	2.4414	3	0.4578	691	0.4119
0.5500	−2	−4.8828	−2	−0.3052	−248	−0.1478
0.6000	1	2.4414	1	0.1526	230	0.1371
0.6500	−2	−4.8828	−3	−0.4578	−427	−0.2545
0.7000	1	2.4414	2	0.3052	276	0.1645
0.7500	−2	−4.8828	−7	−1.0681	−1317	−0.7850
0.8000	1	2.4414	10	1.5259	2482	1.4794
0.8500	−2	−4.8828	−7	−1.0681	−1381	−0.8231
0.9000	−2	−4.8828	−4	−0.6104	−531	−0.3165
0.9500	−2	−4.8828	−5	−0.7629	−958	−0.5710
1.0000	−2	−4.8828	−11	−1.6785	−2490	−1.4842

Note: V_0 in mV

4.4 CONVERSION OF THE MEASURED DATA INTO THE MECHANICAL UNIT

Consider an object vibration of $P = 1$ m/s^2 is measured using an accelerometer of sensitivity, $S = 100$ mV/g. The vibration signal from the sensor is then recorded using a 16-bit DAQ device with input voltage ±5 V. Hence the forward calculation for the recorded voltage V_0 is given by

Sensor sensitivity, $S = 100$ mV/g $= 0.1$ V/g $= (0.1/9.81)$ V/m/s^2
Sensor output, $S_0 = P \times S = 1 \times (0.1/9.81) = 0.0102$ V;
DAQ sensitivity, $S_d = \frac{FSIV}{2^b} = 0.15259$ mV/bin
Number of bins, $B = \frac{S_0}{S_d} = 66.84 = 67$ (integer number above 66)
Recorded voltage, $V_0 B \times S_d = 67 \times 0.15259$ mV $= 0.01022$ V

Vibration Instruments and Measurement Steps

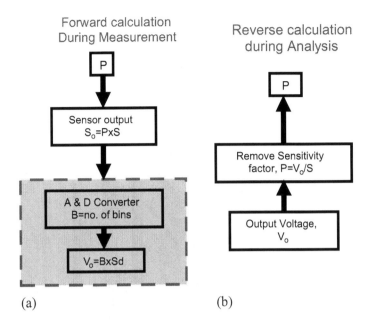

FIGURE 4.31 Forward and reverse calculation procedure (a) forward calculation (b) reverse calculation during analysis.

Similarly, this recorded voltage, V_o must be converted back to the measuring parameter, P during the data analysis. Figure 4.31 gives the process of forward and reverse calculations. Hence the recorded vibration will be $P = V_o/S = 0.01022/0.0102 = 1.002$ m/s² so nearly an error of 0.2%.

4.5 SUMMARY

The chapter summarizes the working principles of different instruments commonly used in the vibration monitoring of machines. The concept of sampling frequency is introduced. The aliasing affect and use of anti-aliasing filter during the data collection using the ADC approach are also explained. The process of digitization of any analog voltage (i.e., ADC) through DAQ device is explained through a simple mathematical concept to ease the understanding. Instruments specifications and their use in the selection process are also highlighted.

REFERENCES

Sinha, Jyoti K. 2002. Health Monitoring Techniques for Rotating Machinery, PhD Thesis, University of Wales Swansea (Swansea University), Swansea, UK, October 2002.

Sinha, Jyoti K. 2015. *Vibration Analysis, Instruments, and Signal Processing*. Boca Raton, FL: CRC Press/Taylor & Francis Group, January 2015. http://www.crcpress.com/product/isbn/9781482231441.

5 Signal Processing

5.1 TIME SIGNAL

The raw signal (amplitude versus time) from any sensor is known as the time domain signal; therefore the collected signal through the DAQ device is nothing but the time domain signal (amplitude versus time of data collection). The measured time domain signals from the different sensors are then to be analyzed to get the required information. There are several signal processing techniques are available to analyze the measured the data to get the meaningful information from the signals (Sinha, 2015). In this chapter, a number of signal processing methods that are commonly used for the vibration signals are discussed.

5.1.1 FILTERS

The anti-aliasing filter discussed in Chapter 4 is used in the DAQ device to remove all frequencies above the Nyquist frequency, f_q from the signals to be collected. This anti-aliasing filter is nothing but the low pass filter that is applied during the process of the data collection. During the signal processing of the recorded data, it may be required to remove few frequencies from the signal to get meaningful information for the machine fault diagnosis. These can be done using following filters:

1. *Low pass (LP) filter:* It is similar to the anti-aliasing filter as shown in Figure 5.1a, but it is generally applied on the recorded/collected/measured data. It removes all frequencies within the measured signal at and above the frequency (f_l). The user can choose the value of the low pass frequency (f_l) depending on the requirements.
2. *High pass (HP) filter:* A HP filter typically removes the frequency content from a signal at and below a particular frequency, f_h. A typical high pass filter at f_h for a signal is shown in Figure 5.1b. In many cases it is required to filter out low frequency noise, DC components, etc. from the recorded/measured signals before further signal processing.
3. *Band pass (BP) filter:* This is a combination of both HP and LP filters. It is typically shown in Figure 5.1c.

The working of these three types of filters is typically illustrated through a signal in Figure 5.2 to aid the clear understanding of filtering process. The signal is collected at the sampling frequency, $f_s = 2000$ Hz. The collected signal contains three frequencies components (50 Hz, 200 Hz, and 500 Hz) at different amplitudes and phases together with low frequencies noise below 1 Hz. It is obvious from Figure 5.2 that the low pass filter at 400 Hz removes the frequency component at 500 Hz from the

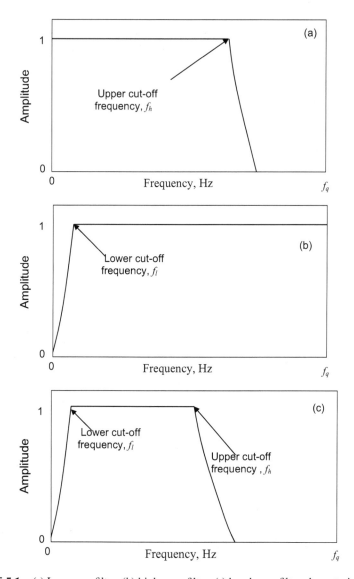

FIGURE 5.1 (a) Low-pass filter, (b) high-pass filter, (c) band-pass filter characteristics.

signal. Similarly the high filter at 100 Hz and the band pass filter between 100 Hz and 400 Hz have removed the frequencies components (low frequencies noise and 50 Hz) and (low frequencies noise, 50 Hz and 500 Hz) respectively from the signal.

5.1.2 Amplitude of Vibration

The overall amplitude of any signal may be specified in any of three ways, namely root mean square (*RMS*), 0 to peak (*pk*) or peak-to-peak (*pk-pk*). Figure 5.3a represents a simple illustration of the RMS, pk and pk-to-pk values of a displacement sine

Signal Processing

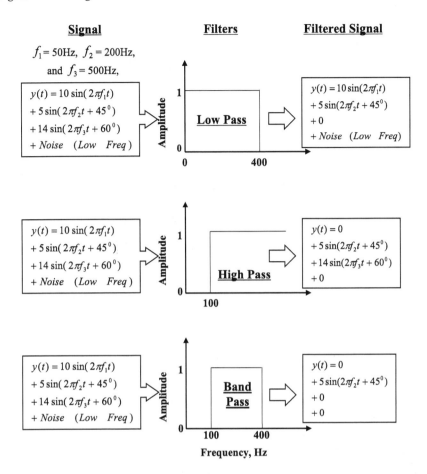

FIGURE 5.2 Working illustrations of different filters.

wave signal at 50 Hz. Figure 5.3b however shows the time waveform of a periodic signal that is non-sinusoidal. The RMS value of any sine wave is always equal to 0.707 times the pk value. If a sinusoidal signal is $y(t) = A \sin \omega t$, then the *RMS* value will be given by $y_{rms} = A/\sqrt{2} = 0.707A$.

However, the measured signals may not be pure sinusoidal waveform (such as that shown in Figure 5.3b), and hence it is better to estimation the RMS value numerically. Let us assume a time domain signal, $y(t)$ which contains "p" data points, represented as $y_i(t_i)$ at time, t_i where $t_i = (i-1)dt$, $i = 1, 2, ..., p$. Then, the *RMS* can be calculated as

$$y_{rms} = \sqrt{\frac{\sum_{i=1}^{p} y_i^2}{p}} \quad (5.1)$$

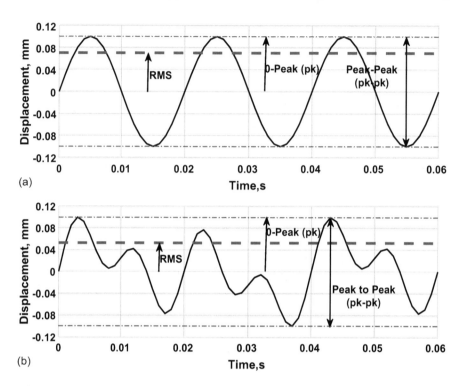

FIGURE 5.3 Time wave forms showing 0-peak, peak-peak, and RMS vibration amplitudes.

The overall vibration value in RMS from the time domain signal seems to present better reflection of any machine vibration as the RMS represents the energy content in the signal and not depend on only 1 or 2 points in the signal for the 0-peak or peak-peak values, respectively. ISO codes also recommend the RMS vibration values for the machine vibration.

5.1.3 Integration of Time Domain Signal

It is often required to convert the measured acceleration signal from an accelerometer into velocity and displacement. Let us assume that the acceleration signal is $a(t) = \ddot{y}(t)$. Generally the exact nature of the measured signal is difficult to express in sine and cosine terms, and so its analytical integration is often difficult. Therefore, it is better to compute numerically by assuming initial velocity, $\dot{y}_0(t = t_0) = 0$, and displacement, $y_0(t = t_0) = 0$. The simple integration process from the acceleration signal to the velocity and displacement is mathematically computed as

Signal Processing

$$\text{The velocity at the time } t_i, v_i = \dot{y}_i = \ddot{y}_i dt + \dot{y}_{i-1} \tag{5.2}$$

and

$$\text{the displacement at the time } t_i, y_i = \dot{y}_i dt + y_{i-1} \tag{5.3}$$

where $t_i = (i-1)dt$, $i = 1, 2, 3,\ldots, p$

For illustration, an example of a displacement signal, $x(t)$ in Equation (5.4) is used. This equation is also differentiated to derive the velocity, $v(t)$ and the acceleration, (t). Expressions for the velocity and acceleration are given in Equations (5.5) and (5.6). Time waveforms for these three signals are shown in Figure 5.4 when the data are sampled using the sampling frequency, $f_s = 100$ Hz.

$$\text{The displacement,} \quad x(t) = 20\sin(3.5\pi t) + 25\cos(6\pi t) \tag{5.4}$$

$$\text{The velocity,} \quad v(t) = 70\pi\cos(3.5\pi t) - 150\pi\sin(6\pi t) \tag{5.5}$$

and

$$\text{the acceleration,} \quad a(t) = -245\pi^2\sin(3.5\pi t) - 900\pi^2\cos(6\pi t) \tag{5.6}$$

Hence the time step, $dt = 1/f_s = 0.01$s is used to compute the velocity and displacement data are also computed from the acceleration signal, $a(t)$ using the numerical integration scheme in Equations (5.3) and (5.4). The computed displacement and velocity data are also plotted in Figure 5.4 for direct comparison. It can be seen in Figure 5.4

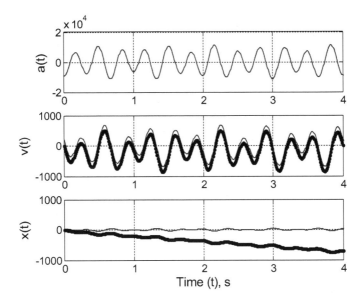

FIGURE 5.4 Original acceleration, velocity and displacement signals (solid line) and comparison with velocity and displacement signals (solid line with dots) computed from the original acceleration signals using the numerical integration.

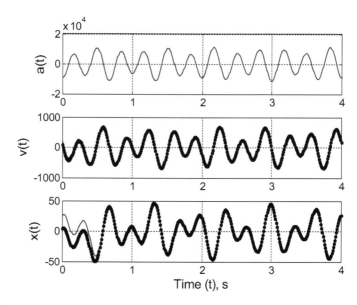

FIGURE 5.5 Original and integrated signals—velocity and displacement after high pass filter at 1 Hz showing promising results.

that the signals obtained from the numerical integration method show significant deviation from the original signals. This only indicates that such integration introduces some error due to the fact that the initial conditions are generally not known that results into DC offset in the integrated signals due to the integration constant.

These DC components in the integrated signals (velocity and displacement) can be considered as the very low frequency components in the signal. Hence, it is always recommended that the integrated signals must be high-pass filtered at a very low frequency. Here the integrated signals—velocity and displacement signals—are high-pass filtered at 1 Hz. It can be seen that the filtered signals shown in Figure 5.5 are almost identical to the original signals after 0.5 s. Thus, it is also recommended to remove some initial data after integration to get the reliable data. For the present example, the data up to 0.5 s must be removed from the integrated data sets.

5.1.4 Statistical Parameters

It is often important to understand the nature of the signal whether it is sine, random or impulsive. It can be examined through the statistical parameters—Crest Factor and Kurtosis.

1. *Crest Factor (CF)*: This is a non-dimensional parameter for any signal, which is defined as the ratio of the *peak value to the RMS value*. For sine wave, it is always going to be equal to 1.414 or $\sqrt{2}$ because $y_{rms} = y_{peak}/\sqrt{2}$.

$$CF = \frac{y_{peak}}{y_{rms}} \quad (5.7)$$

Signal Processing

2. *Kurtosis (Ku)*: This parameter is defined in statistics as the normalized fourth order moment of the time domain signal. The definition of the kth order central moment for any time domain signal (say, $y(t)$) is given by

$$M_k = \frac{1}{p} \sum_{i=1}^{p} (y_i - \bar{y})^k \tag{5.8}$$

where $y_i = y(t_i)$, $\bar{y}$ is the mean value of the data set, $y(t)$, and $i = 1, 2, 3, \ldots, p$. The normalized fourth order moment, *Kurtosis (Ku)*, is computed as

$$Ku = \frac{M_4}{(M_2)^2} \tag{5.9}$$

An example of a sine signal of frequency 1 Hz shown in Figure 5.6 is considered to demonstrate the computation of the CF and Ku. The signal is sampled at the sampling frequency, $f_s = 20$ Hz (i.e., the time step, $dt = 1/f_s = 0.05$ s). The dots on the sine wave are the sampled data. The measured (sampled) data of the sine signal and the calculation of the CF and Ku are listed in Table 5.1. As expected the crest factor (CF) is close to $\sqrt{2}$ for a sine signal. It can be exactly equal to $\sqrt{2}$ if the signal is sampled/collected at higher sampling frequency. Similarly the kurtosis (Ku) should have value close to 1.5 for a sine signal.

5.1.5 Comparison between CF and Kurtosis

For illustration, 2 scenarios are considered here.

Scenario 1: Two typical signals are shown in Figure 5.7. One signal is a pure random signal as shown in Figure 5.7a. Another signal is shown in Figure 5.7b which is again the pure random signal in Figure 5.7a but added with few impulsive signals. The calculated values are ($Ku = 3.038$ and $CF = 3.58$) and ($Ku = 6.75$ and $CF = 7.54$) for the pure random and the random plus impulsive signals in Figure 5.7, respectively.

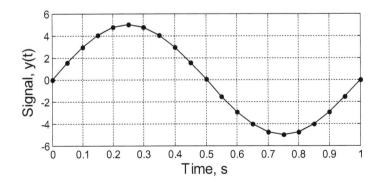

FIGURE 5.6 A typical signal, $y(t)$.

TABLE 5.1
Calculation of CF and Ku for a Sine Signal

S. No.	t, s	$y(t)$ or y_i		y_i^2	$\left(y_i - \bar{y}\right)^2$	$\left(y_i - \bar{y}\right)^4$
1	0	y_1	0.0	0.0	0.0	0.0
2	0.05	y_2	1.5451	2.3873	2.3873	5.6991
3	0.10	y_3	2.9389	8.6373	8.6373	74.6027
4	0.15	y_4	4.0451	16.3627	16.3627	267.7384
5	0.20	y_5	4.7553	22.6127	22.6127	511.3348
6	0.25	y_6	5.0000	25.0000	25.0000	625.0000
7	0.30	y_7	4.7553	22.6127	22.6127	511.3348
8	0.35	y_8	4.0451	16.3627	16.3627	267.7384
9	0.40	y_9	2.9389	8.6373	8.6373	74.6027
10	0.45	y_{10}	1.5451	2.3873	2.3873	5.6991
11	0.50	y_{11}	0.0	0.0	0.0	0.0
12	0.55	y_{12}	−1.5451	2.3873	2.3873	5.6991
13	0.60	y_{13}	−2.9389	8.6373	8.6373	74.6027
14	0.65	y_{14}	−4.0451	16.3627	16.3627	267.7384
15	0.70	y_{15}	−4.7553	22.6127	22.6127	511.3348
16	0.75	y_{16}	−5.0000	25.0000	25.0000	625.0000
17	0.80	y_{17}	−4.7553	22.6127	22.6127	511.3348
18	0.85	y_{18}	−4.0451	16.3627	16.3627	267.7384
19	0.90	y_{19}	−2.9389	8.6373	8.6373	74.6027
20	0.95	y_{20}	−1.5451	2.3873	2.3873	5.6991
21	1.00	y_{21}	0.0	0.0	0.0	0.0

$$\text{Mean, } \bar{y} = 0$$
$$\text{Peak} = 5$$

$$\text{Sum, } \sum_{i=1}^{21} y_i^2 = 250$$

$$y_{RMS} = \sqrt{\frac{\sum_{i=1}^{21} y_i^2}{21}} = 3.4503$$

$$\sum_{i=1}^{21}\left(y_i - \bar{y}\right)^2 = 250$$

$$M_2 = \frac{\sum_{i=1}^{21}\left(y_i - \bar{y}\right)^2}{21} = 11.9048$$

$$\sum_{i=1}^{21}\left(y_i - \bar{y}\right)^4 = 4687.50$$

$$M_4 = \frac{\sum_{i=1}^{21}\left(y_i - \bar{y}\right)^4}{21} = 223.2143$$

$$\text{Crest factor, } CF = \frac{y_{peak}}{y_{rms}} = \mathbf{1.4491}$$

$$\text{Kurtosis, } ku = \frac{M_4}{(M_2)^2} = \mathbf{1.575}$$

It is obvious from the values that the crest factor and kurtosis are close to 3 for the random signal but much higher values in case of impulsive signal.

Figure 5.8 explains the reason for higher kurtosis value in case of impulsive signal. The histogram of the impulsive signal shows much longer tails on either side of the signal amplitude distribution compared to the random data. This results into much bigger value of the fourth order central moment for the signal. Hence the higher value for the kurtosis in case of the impulsive signal.

Scenario 2: Typical three types of signals shown in Figure 5.9 are considered here. The signal $a_1(t)$ is a pure sine wave at 5 Hz, the signal $a_2(t)$ is a normally

Signal Processing

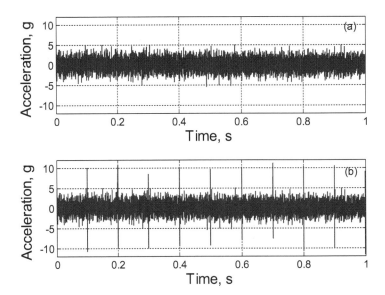

FIGURE 5.7 Typical signals (a) pure random ($Ku = 3.038$, $CF = 3.58$), and (b) random plus impulsive ($Ku = 6.75$, $CF = 7.54$).

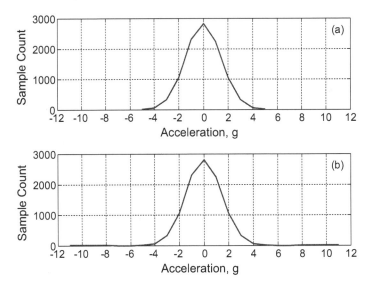

FIGURE 5.8 Typical histogram plots of the signals in Figure 5.7 (a) Pure random and (b) random plus impulsive.

distributed random data and the signal $a_3(t)$ is the transient vibration data due to the impulsive loading. The peak, RMS, CF and Ku are computed for these data and are listed in Table 5.2 for comparison. The peak value is kept same for all three signals so that easy comparison of different estimated parameters for the three signals can be done. It can be seen that the CF and Ku both increased significantly for both the

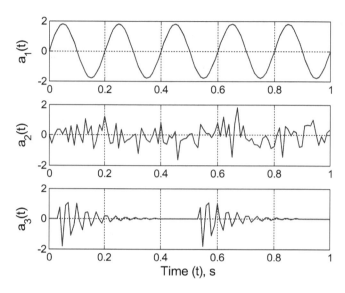

FIGURE 5.9 Three signal waveforms—sine, random, and transient (impulse response type).

TABLE 5.2
Comparison of Different Estimated Parameters for the Three Signals—Sine, Random and Impulsive (Transient)

Signals	Mean	Peak	RMS	CF	Ku
Sine, $a_1(t)$	1.892e-16	1.80	1.2728	1.4142	1.515
Random, $a_2(t)$	−3.851e-2	1.80	0.6108	2.9469	3.2563
Impulsive, $a_3(t)$	2.737e-5	1.80	0.4400	4.0905	8.4486

random and transient data compared to the sine wave data. The Kurtosis (Ku) shows sharp increase in value for the impulsive signal compared to other two signals for his example.

The kurtosis may be a better and reliable parameter to understand the nature of signals as it involves all data points within the signal for the computation of the kurtosis. However, the crest factor is simple but significantly dependent on a just one peak value of the signal. If the peak data in the signal is due to some measurement error then the CF value is going to be meaningless.

5.2 FOURIER TRANSFORMATION (FT)

It is important to convert time domain waveform signal into frequency domain in order to ascertain the different frequency contents within a signal. This helps to understand the physical behavior of the measuring parameters within a system. The amplitude vs. frequency plot of any signal is known as the spectrum of the signal. The computation is explained through a sine wave signal in following sections.

5.2.1 Example 5.1: A Sine Wave Signal

A sine wave signal shown in Figure 5.10 is considered here. It is collected at the sampling frequency, $f_s = 1024$ Hz. The zoomed view of the signal is shown in Figure 5.11 clearly indicates that the signal is a pure sine wave with time period equal to 0.01 s (i.e., 100 Hz).

Since the signal shown in Figures 5.10 and 5.11 is a pure sine wave of frequency 100 Hz, with amplitude equal to 1. Hence the spectrum should have ideally a peak at 100 Hz with amplitude equal to 1. The ideal spectrum for this signal is shown in Figure 5.12. But in practice the measured signals from different sensors on a machine may not be pure sine wave signals and hence it is difficult to estimate their spectrum plots by visual means.

It is also known that any signals can be represented by the Fourier series. Once the expression for Fourier series is known then spectrum can be generated directly. However it is difficult to write the expression for the Fourier series for the measured signals in most cases. Hence an alternative method is developed and used based on the concept of Fourier series which is known as Fourier Transformation (FT).

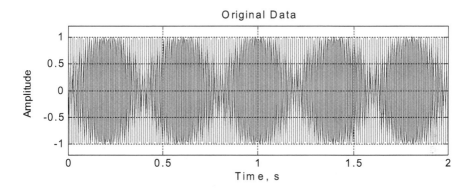

FIGURE 5.10 A typical 100 Hz sinusoidal signal for the FT.

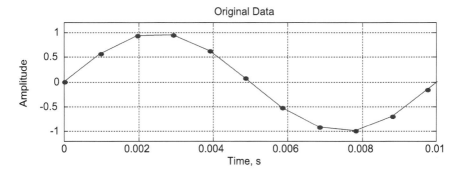

FIGURE 5.11 Zoom view of Figure 5.10 showing just a complete cycle.

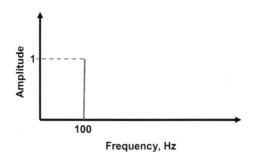

FIGURE 5.12 Expected spectrum of the sine wave shown in Figure 5.10.

5.2.2 Steps Involved for the Computation of FT

Following steps are generally needed for the FT:

- The actual time period of the measured signal may not be known. Hence it is essential to define artificially the time period (T) of the signal for the FT calculation.
- Frequency resolution, $df = 1/T$.
- Maximum frequency can be computed up to Nyquist frequency, f_q but it is better to consider the spectrum up to the useful frequency limit, $f_u = f_s/2.56$ due the anti-aliasing filter used in the data collection.

Fourier transformation (FT) is a tool that is used for converting digital data in a time domain into the frequency domain. The process is also known as the Digital FT (DFT). It can be calculated using the following formula:

$$Y(kdf) = \frac{2}{N} \sum_{p=1}^{N} y(t_p) e^{-j\frac{2\pi(p-1)k}{N}} \quad (5.10)$$

where $Y(kdf) = Y(f_k)$ is the FT at the frequency, $f_k = kdf$ for the time domain signal, $y(t)$. $y_p = y(t_p)$ is the amplitude of the signal, $y(t)$ at time, $t_p = (p-1)dt$. N is the number of data points of the signal, $y(t)$ used for the FT.

The variables $k = 0, 1, 2, ..., (N/2 - 1)$ and $p = 1, 2, ..., N$. The frequency resolution, $df = 1/T$, where T is the pseudo (artificial) time period chosen for the signal for the FT and is estimated as $T = Ndt$, where $dt = 1/f_s$ is the time between two consecutive data in the signal, $y(t)$.

The number of data points (N) used for the FT calculation should be equal to 2^n, where n is a positive integer number. The selection of number of data points N (or n) from the time domain signal data is totally the choice of the user, as this defines the artificial time period, T and the frequency resolution, df for the calculated spectrum. Both frequency and amplitude axes from any time domain signal can be computed using Equation (5.10) which is typically shown in Figure 5.13. The computed FT value, $Y(f_k)$ at the frequency, f_k is a complex quantity (i.e., $a_k + jb_k$). Hence, the

Signal Processing

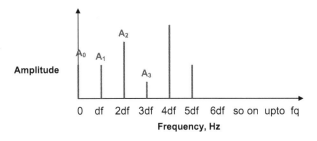

FIGURE 5.13 Spectrum plot: a graphical representation of the computed amplitude and frequency through FT using Equation (5.10).

amplitude and phase at a frequency, f_k in a signal can be calculated as $\sqrt{a_k^2 + b_k^2}$ and $\tan^{-1}\left(\frac{b_k}{a_k}\right)$ respectively. But only amplitudes are shown in the spectrum at different frequencies. The computed phase value at each frequency is generally ignored if there is no definite reference for the data collection.

It is also important to note that the spectrum can be computed up to Nyquist frequency, f_q but it is better to consider the spectrum only up to the useful frequency limit, $f_u = f_s/2.56$. Hence, the number of frequency lines (or FFT lines) in the computed spectrum will be $N_L = \frac{N}{2.56}$ starting from frequency, df, $2df$, $3df$, up to f_u (0 Hz is not included in the FFT lines counting). $N = 2^n$ is also known the FFT points of the signal required for the FT. The number of FFT lines, N_L is always a positive integer number.

Earlier days these calculations were performed through the electronics circuits when the computers and computational programming were not common and readily available. Those days this DFT approach was known as the Fast FT (FFT) and instruments performing this FFT as the FFT analyzers.

Table 5.3 gives the summary of different parameters used in the FFT for a time domain signal data shown in Figure 5.10 collected at the sampling rate $f_s = 1024$ Hz

TABLE 5.3
Parameters for FT Analysis

n	$N = 2^n$	$T = Ndt$, s	$df = \frac{1}{T}$, Hz	$f_u = \frac{f_s}{2.56}$, Hz	$N_L = \frac{N}{2.56}$
1	2	0.001953	512	400	0
2	4	0.003906	256	400	1
3	8	0.007812	128	400	3
4	16	0.015625	64	400	6
5	32	0.03125	32	400	12
6	64	0.0625	16	400	25
7	128	0.125	8	400	50
8	256	0.25	4	400	100
9	512	0.50	2	400	200
10	1024	1.0	1	400	400

(samples/s) and the time step, $dt = 1/f_s = 0.97656$ ms. Table provides the comparison of the spectrum analysis (i.e., FT) starting from $n = 1$ to 10.

The spectrum is now computed for the time domain sine wave signal in Figure 5.10 using Equation (5.10) but for three scenarios of Table 5.3.

Scenario (a): Only a number of data points within the signal, $N = 2^n = 2^6 = 64$ is considered for the FT analysis. The computed spectrum is shown in Figure 5.14a. The "dots" on the spectrum indicate the computed FT values i.e., FFT lines, $N_L = \frac{N}{2.56} = \frac{64}{2.56} = 25$ frequencies from 16 Hz to 400 Hz at a step to 16 Hz plus 0 Hz. The amplitude of 1 at 100 Hz for the signal

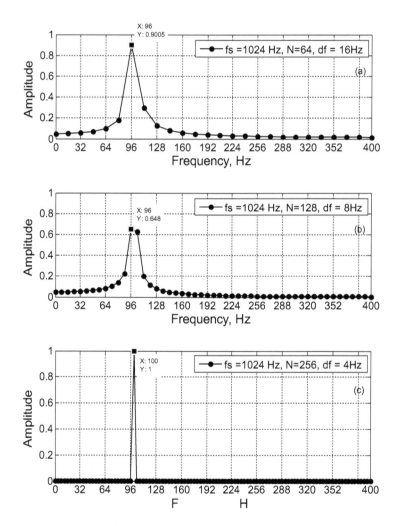

FIGURE 5.14 Estimated spectrum plots using the FT approach for the sine wave signal in Figure 5.10, (a) $N = 64$ and $df = 16$ Hz, (b) $N = 128$ and $df = 8$ Hz, (c) $N = 256$ and $df = 4$ Hz.

Signal Processing

(see Figure 5.12) is appearing as the amplitude of 0.9005 at 96 Hz due to the frequency resolution of $df = \frac{1}{T} = 16$ Hz is used in the spectrum analysis.

Scenario (b): Same signal shown in Figure 5.10, but the number of data points within the signal, $N = 2^n = 2^7 = 128$ is considered for the FT analysis. The computed spectrum is shown in Figure 5.14b. Now the frequency resolution is 8 Hz and FFT lines, $N_L = \frac{N}{2.56} = \frac{128}{2.56} = 50$ frequencies represented by "dots" on the spectrum shown in Figure 5.14b from 8 Hz to 400 Hz at a step to 8 Hz plus 0 Hz. Since the frequency resolution of 8 Hz is not the multiple of 100 Hz in the signal, the peaks are appearing at 96 Hz and 104 Hz instead of a peak at 100 Hz.

Scenario (c): Once again same signal shown in Figure 5.10 but $N = 2^n = 2^8 = 256$ data points are considered for the FT analysis. Now the frequency resolution becomes equal to 4 Hz, which is now multiple of 100 Hz. Hence the computed spectrum shown in Figure 5.14c is clearly showing a peak at 100 Hz with 1 amplitude exactly same as the ideal spectrum shown in Figure 5.12 for the sine wave signal considered here.

5.2.3 IMPORTANCE OF FREQUENCY RESOLUTION IN SPECTRUM ANALYSIS

Earlier Example 5.1 and its spectrum analysis clearly highlight the importance of appropriate selection of the frequency resolution in the spectrum. To further strengthen this observation, another signal, which consists of 2 frequencies, 99 Hz and 101 Hz and both having amplitude of 1 is considered here. The signal is sampled at the sampling frequency, $f_s = 1024$ Hz.

Initially the spectrum analysis is carried out using the FT parameters, $N = 2^n = 2^8 = 256$ data points from the signal, $T = N\,dt = 0.25$ s, $df = \frac{1}{T} = 4$ Hz, $N_L = \frac{N}{2.56} = \frac{256}{2.56} = 100$ frequency lines and $f_u = \frac{f_s}{2.56} = \frac{1024}{2.56} = 400$ Hz. The computed spectrum is shown in Figure 5.15. A frequency peak at 100 Hz at amplitude of 1.273 is seen only in the spectrum instead of 99 and 101 Hz due to poor frequency resolution, $df = 4$ Hz. Hence this spectrum is not correct and likely to leads to incorrect interpretation.

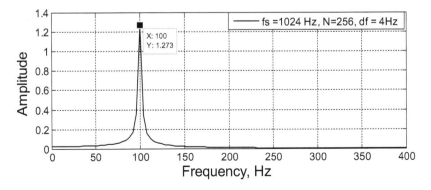

FIGURE 5.15 Estimated spectrum plot—peaks at 99 Hz and 101 Hz appearing as a peak at 100 Hz when $N = 256$ and $df = 4$ Hz used.

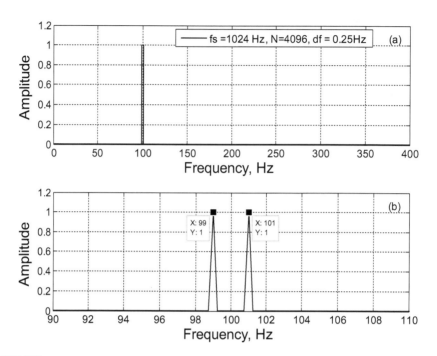

FIGURE 5.16 Estimated spectrum plots: (a) peaks at 99 Hz and 101 Hz clearly seen when $N = 2048$ and $df = 0.50$ Hz used, (b) zoomed view in frequency range 90 Hz to 110 Hz.

The spectrum analysis is again carried out but using the FT parameters, $N = 2^n = 2^{11} = 2048$ data points from the signal, $T = N\,dt = 2$s, $df = \frac{1}{T} = 0.5$ Hz, $N_L = \frac{N}{2.56} = \frac{2048}{2.56} = 800$ frequency lines and $f_u = \frac{f_s}{2.56} = \frac{1024}{2.56} = 400$ Hz. The computed spectrum is shown in Figure 5.16a and its zoomed view between 90 and 110 Hz in Figure 5.16b. The computed spectrum is clearly showing both 99 Hz and 101 Hz frequency peaks with their amplitudes equal to 1. Hence this spectrum is correct representation of the signal.

Therefore it is always important to understand the measured signals from any machines and then select the appropriate FT parameters to ensure the spectrum analysis adequately representing the time domain signals.

5.2.4 Leakage

A time domain signal of a sine wave of 5 Hz is considered here. The signal is sampled at the sampling frequency, $f_s = 930$ Hz up to the number of the data points, $N = 2^{10} = 1024$. A collected signal of the segment of size $N = 2^{10} = 1024$ of time length, $T = N\,dt = 1.101$ s at the sampling frequency, $f_s = 930$ Hz, shown in Figure 5.17, is used for the computation of the FT and hence its spectrum.

The time period of the signal (T_s) is equal to 0.2 s for the 5 Hz sinusoidal signal. But the selected segment size (N) of the segment time length, T is not the integral multiple of the time period, T_s. The extra data in the selected segment size,

Signal Processing

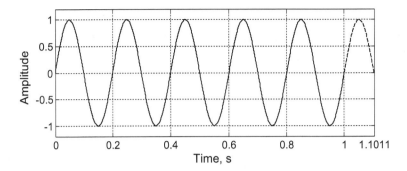

FIGURE 5.17 A typical segment size $T = Ndt$ of a sine waveform.

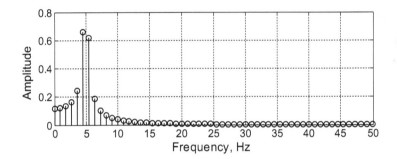

FIGURE 5.18 Frequency peak with side peaks indicate the leakage in the FT.

$T = Ndt$ for the FT analysis is deliberately shown as the dash line from 1 to 1.1011 s in Figure 5.17. This may impact the spectrum feature calculated using Equation (5.10).

Ideally, the spectrum should have a frequency peak at 5 Hz with amplitude equals to 1 for the time domain signal shown in Figure 5.17. However the artificial selection of time period, T for the segment of the signal for the FT analysis has resulted in the spectrum shown in Figure 5.18. The spectrum is shown up to 50 Hz only to clearly observe the features around 5 Hz. The appearance of small side peaks on either side of 5 Hz is often due to the fact that the time length of the segment may not be equals to or integral multiple of the actual time period of the signal. This effect is known as the leakage in the spectrum computation.

5.2.5 Window Functions

Leakage in the spectrum is a serious problem and may results in misleading vibration analysis and conclusion. Hence this error must be avoided during signal processing. It is well known that the selection of the segment size, $T = Ndt$ is totally dependent on the number of data points, $N = 2^n$ chosen by the users during the spectrum analysis in their commercially analyzers or computation codes. This selection may not match with the time period of the signal and in most of cases, the time period of the measured signals are often difficult to know. The use of the window function is found to overcome the leakage problem in signal processing. A number of window functions are suggested and used in practice for this purpose. A few different types of windows are listed here:

1. Flat top window/rectangular window—This means there is no window used for the FT calculation and Equation (5.10) is directly used for the FT and spectrum calculation.
2. Gaussian window
3. Hanning window
4. Hamming window

The typical shapes of these windows are shown in Figure 5.19 for the segment size, $N = 2^{10} = 1024$. Note that the shapes remain the same for these window functions irrespective of the segment size, N. The idea is to make the selected segment of size, $T = N\mathrm{dt}$ of any signal into periodic signal artificially, by using either of these windows to reduce the leakage problem in the spectrum using FT approach. Let us assume that the window function is denoted by w_p, where $p = 1, 2, 3, ..., N$. Hence the segment of the time domain signal, $y(t)$ is modified as $y_{w,p} = y_p.w_p$ using the dot product of the vectors of the time domain segment of the signal and the window function. The window function, w_p can be estimated using equations

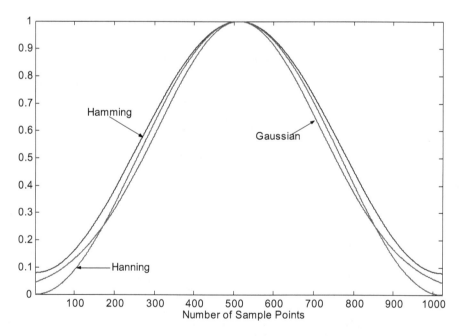

FIGURE 5.19 Typical view of a few different windows for the segment size, $N = 1024$.

Signal Processing

$$w_p = 0.5\left(1-\cos\left(\frac{2\pi(p-1)}{(N-1)}\right)\right), \quad w_p = \alpha - \beta\cos\left(\frac{2\pi(p-1)}{(N-1)}\right),$$

(5.11)

(where $\alpha = 0.54$ and $\beta = 1-\alpha$) and $w_p = e^{-\frac{1}{2}\left(\frac{(p-1)-\frac{(N-1)}{2}}{0.4\frac{(N-1)}{2}}\right)^2}$

for the Hanning, Hamming, and Gaussian windows, respectively (Bendat and Piersol, 1980, Oppenheim and Schafer, 1989). The modified time domain signal in Figure 5.17 using Hanning window is shown in Figure 5.20 together with the Hanning window. The use of a window makes the amplitude nearly zero at the start and end of the chosen segment of the time domain signal. Now the signal segment of N data looks more like a periodic signal. The FT calculation on this modified signal is likely to satisfy the periodic condition of the Fourier series and hence it may help in reducing the leakage effect in the spectrum. The use of the window function affects the signal amplitude, hence the FT calculation in Equation (5.10) is modified as

$$Y(kdf) = \frac{1}{C_w}\sum_{p=1}^{N} y_{w,p} e^{-j\frac{2\pi(p-1)k}{N}}$$

(5.12)

where C_w is a constant that compensate the effect of the window function on the amplitude of the computed spectrum. C_w is estimated as

$$C_w = \left(\frac{\sqrt{\sum_{p=1}^{N} w_p^2}}{\sum_{p=1}^{N} w_p / N}\right)$$

(5.13)

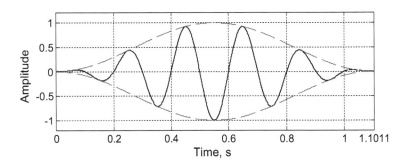

FIGURE 5.20 Sine waveform of Figure 5.17 with Hanning window.

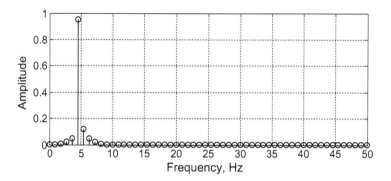

FIGURE 5.21 Spectrum with reduced leakage.

The spectrum estimated for the modified signal, shown in Figure 5.20 using Equation (5.12) is shown in Figure 5.21. It is clearly seen that the leakage effect is reduced significantly and a distinct peak at 5 Hz with amplitude of nearly 1 is observed.

5.3 COMPUTATION OF POWER SPECTRAL DENSITY (PSD)

It is briefly discussed earlier in Section 5.2.2 that the FT analysis estimates both amplitude and phase at each frequency for any signal up to the Nyquist frequency but only amplitudes are shown in the spectrum. The reason for this further explained in this section through the averaging process that includes the concept for the power spectrum density (PSD) of a signal.

5.3.1 Averaging Process

It is well known that the measured signal generally contains noise. It is also known that the FT computation is done through an approximate method based on the concept of Fourier series, which depends heavily on the artificially selected time period, $T = Ndt$. A typical time domain signal is shown in Figure 5.22. The time length of the signal is sufficient long to divide the signal into a number of equal segments of each segment size, $N = 2^n$. The FT analysis can then applied to all segments one by one. This process will give a number of spectra for a signal, but each spectrum per segment may contain random spurious frequency peaks in addition to the real frequency peaks related to the signal either due to random noise in the signal or artificial time period, T due to selection of segment size, N or both.

The averaging of all these spectra may minimize these errors but the averaging process is not straight forward. It is because the FT analysis computes a complex number at each frequency. This means the FT computes both amplitude and phase at each frequency but phase angle at a frequency may differ significantly from one segment to another segment of a signal. It is because the selection of starting point for each segment within a signal is completely random which is not related to the actual time period of signal or a common reference starting point for all segments.

Signal Processing

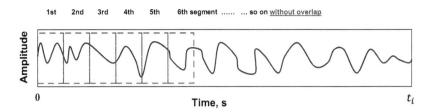

FIGURE 5.22 Process of dividing the time domain signal into a number of segments of size, N.

The phase information may not be useful and thus it is ignored in the spectrum; therefore, it is important to convert the complex number of FT at each frequency into real number before averaging process. The following approach is suggested in practice for the averaging. This is known as the power spectral density (PSD), $S_{yy}(f_k)$ at the frequency, f_k for the time domain signal, $y(t)$.

$$S_{yy}(f_k) = \frac{\sum_{i=1}^{n_s} Y_i(f_k) Y_i^*(f_k)}{n_s} \quad (5.14)$$

where $Y_i(f_k)$ and $Y_i^*(f_k)$ are the FT and its complex conjugate at frequency, f_k for the i th segment of the signal $y(t)$. "n_s" is the number of segments of the signal, $y(t)$ used for averaging. The multiplication of the FT and its complex conjugate for any segment within a signal creates a real number and hence, addition and averaging is possible. The component $Y_i(f_k)Y_i^*(f_k)$ makes magnitude at each frequency as the power of amplitude (i.e., "Amplitude^2") of the signal, $y(t)$. Hence the computed $S_{yy}(f_k)$ is call the PSD of the signal, $y(t)$. Therefore, the amplitude of the spectrum is then estimated as

$$\text{Amplitude of the spectrum} = \sqrt{S_{yy}(f_k)} \quad (5.15)$$

To aid the advantage of the averaging process, the sine wave signal in Figure 5.10 of Example 5.1 is considered again. But now it is fully dominated by the noise. The actual signal is still same i.e., sine wave of frequency 100 Hz with amplitude equal to 1 but now with random noise. The data is again collected at $f_s = 1024$ Hz and is shown in Figure 5.23. The signal amplitude is higher than 1 due to the presence of noise in the signal. The zoom view of the signal in time 0–0.01 s is also shown in Figure 5.24 which is clearly showing no evidence of 100 Hz in the signal.

Now, the 10 s data is divided into 20 equal segments of size $N = 512$ and $T = N dt = 0.5$ s for FT analysis to compute the averaged spectrum. The computed spectrum for a segment (without average or 0 average) and the averaged spectrum of 20 segments are shown in Figure 5.25. It is obvious from Figure 5.25b that the averaged spectrum almost accurately detects the presence of the sine wave at 100 Hz with amplitude close to 1. This example clearly illustrates the advantage of the averaging process in the spectrum computation.

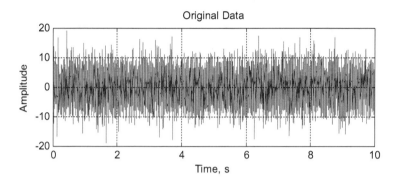

FIGURE 5.23 Collected noisy sine wave data of 100 Hz for Example 5.1.

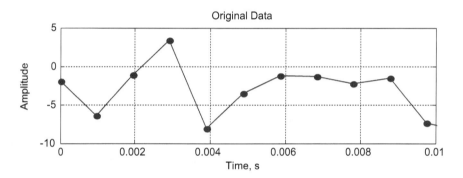

FIGURE 5.24 Zoomed view of the data in Figure 5.23.

5.3.2 Concept of Overlap in the Averaging Process

It is obvious from the earlier discussion and the example that the averaging process in the computation of spectrum is useful to minimize the effect of the noise content in the signals and the artificial selection of the signal time period in terms of the segment size for the FT. A larger of number may be better, however, in many occasion the recorded data length may not be long enough to have a large number of average. Hence, a large number of averaging for the limited short time domain data is possible in the following two ways:

1. Select small segment size, N to increase the number of average but this may impact the frequency resolution for the computed spectrum. This may not be preferred in many vibration analysis applications.
2. Another approach is to keep the segment size good enough to get the required frequency resolution in the spectrum and then increase the number of the averages by the overlap process. The overlap process in averaging is now explained here.

Signal Processing

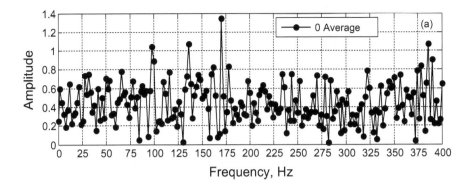

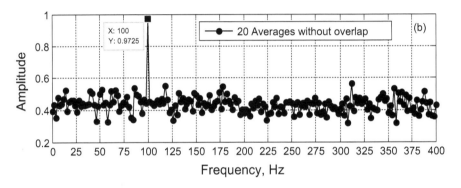

FIGURE 5.25 Spectrum plots, (a) without average showing no clear peak at 100 Hz, (b) with 20 averages showing a clear peak at 100 Hz.

The signal in Figure 5.23 is considered here again. Total length of the signal is 10 s with the sampling frequency, f_s =1024 Hz. This means that signal contains $10 \times 1024 = 10240$ data points. Hence, the signal was divided into ten segments with a segment size, $N = 2^{10} = 1024$ and $T = Ndt = 1$ s for the FT calculation and 10 averages. The division of the ten segments for the signal is illustrated in Table 5.4 for clear understanding. This process is called averaging without overlap, however the overlap process can increase the number of averages. For example, 50% overlap can increase the number of averaging to 19. The 50% overlap means that the first segment contains the data from 0 s (1st data point) to 1 s (1024th data point), then second segment starts at 0.5 s (513rd point) to 1.5 s (1536th point) and so on. This means that only 50% of the data points are dropped from the previous segment and the new 50% data are added to the next segment. This is also demonstrated in Table 5.4. More % in overlap can further increase the number of averages.

TABLE 5.4
Key Elements in the Averaging Process

	Average Without Overlap			Average Without 50% Overlap		
Segment No	Segment Size, N	Start Point	End Point	Segment size, N	Start Point	End Point
1	1024	1	1024	1024	1	1024
2	1024	1025	2048	1024	513	1536
3	1024	2049	3072	1024	1025	2048
4	1024	3073	4096	1024	1537	2560
5	1024	4097	5120	1024	2049	3072
6	1024	5121	6144	1024	2561	3584
7	1024	6145	7168	1024	3073	4096
8	1024	7169	8192	1024	3585	4608
9	1024	8193	9216	1024	4097	5120
10	1024	9217	10240	1024	4609	5632
11	–	–	–	1024	5121	6144
12	–	–	–	1024	5633	6656
13	–	–	–	1024	6145	7168
14	–	–	–	1024	6657	7680
15	–	–	–	1024	7169	8192
16	–	–	–	1024	7681	8704
17	–	–	–	1024	8193	9216
18	–	–	–	1024	8705	9728
19	–	–	–	1024	9217	10240

5.3.3 EXAMPLE 5.2: AN EXPERIMENTAL RIG

An experimental rig consists of two shafts connected through a solid coupling and supported through four ball bearings as shown in Figure 4.26a. It is driven by a motor. Typical measured vibration acceleration spectrum plots at a bearing housing of the rig are also shown in Figure 5.26b and c when the rig was operation at 2400 RPM (40 Hz). Both spectra clearly show the frequency peaks at the RPM and its multiples; however, the spectrum shown in Figure 5.26b (i.e., without any averaging) is definitely noisy compared to the spectrum with 60 averages in Figure 5.26c.

5.3.4 EXAMPLE 5.3: AN INDUSTRIAL BLOWER

A simple schematic of an industrial blower is shown in Figure 5.27a (Elbhbah et al., 2016). The vibration measurements were carried out when the blower speed was 718 RPM (11.96 Hz). Here again, a similar observation is made between the

Signal Processing

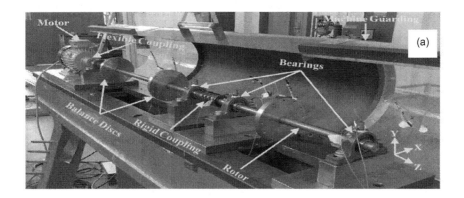

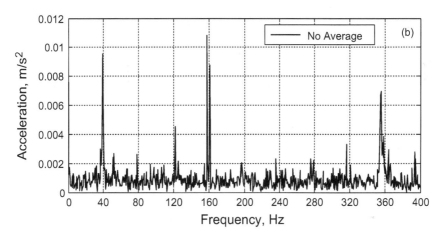

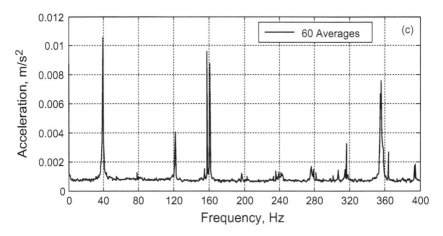

FIGURE 5.26 (a) An experimental rotating rig, (b) measured spectrum (no average), (c) measured spectrum (60 averages).

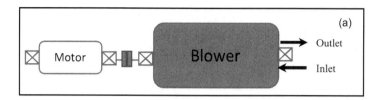

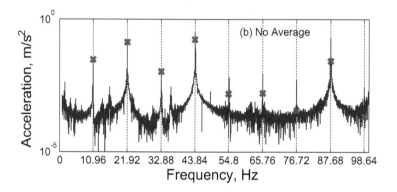

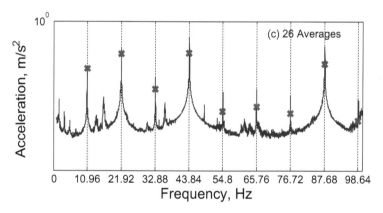

FIGURE 5.27 (a) Industrial blower, (b) measured spectrum (no average), (c) measured spectrum (26 averages).

spectrum plots (with and without averaging) for the measured vibration acceleration on an industrial blower. These spectrum plots are shown in Figure 5.27b and c for the comparison. Once again the averaged spectrum is much better than the spectrum without any average; therefore, it is always recommended to have averaged spectrum if possible.

Signal Processing

5.4 CONVERSION OF ACCELERATION SPECTRUM TO DISPLACEMENT SPECTRUM AND VICE VERSA

The numerical integration and differentiation of the time domain data are possible but may have some error due to no information about the initial conditions. It is typically discussed and demonstrated in Section 5.1.3. However, the integration and differentiation processes are straightforward calculation in the frequency domain to convert the acceleration spectrum to velocity and displacement spectra and vice versa. Typical conversion factors at each frequency in the spectrum are shown in Table 5.5 for the spectrum integration and differentiation. It is important to note that these factors must be ignored at the frequency of 0 Hz. Most of the commercial data or FFT analyzers, the integration functions are represented by $1/j\omega_k$ (for single integration), $1/-\omega_k^2$ (for double integration). Similarly, $j\omega_k$ and $-\omega_k^2$ represents the single and double differentiations respective of the spectrum data. These functions are useful to convert the displacement spectrum to the velocity and acceleration spectra respectively.

Typical examples are shown in Figure 5.28. Figure 5.28a–c are showing the differentiation of the displacement spectrum to the velocity spectrum and then the acceleration spectrum using the estimation methods listed in Table 5.5.

Similarly if the Figure 5.28 is viewed from the Figure 5.28c to Figure 5.28b to Figure 5.28a then it becomes an example of the integration from the acceleration spectrum to the velocity spectrum and then the displacement spectrum respectively. This is very common practice for the industrial vibration measurements as the accelerometers are generally used to measure vibration acceleration of machines and structures.

TABLE 5.5

Conversion Methods (Integration and Differentiation Methods) for the Frequency x Domain Data

Frequency	Displacement spectrum	to	Velocity	or	Acceleration spectrum
f_k	Y_k		$(2\pi f_k)Y_k$		$(2\pi f_k)^2 Y_k$
Frequency	Velocity spectrum	to	Displacement	or	Acceleration spectrum
f_k	V_k		$\dfrac{V_k}{(2\pi f_k)}$		$(2\pi f_k)V_k$
Frequency	Acceleration spectrum	to	Displacement	or	Velocity spectrum
f_k	A_k		$\dfrac{A_k}{(2\pi f_k)^2}$		$\dfrac{A_k}{(2\pi f_k)}$

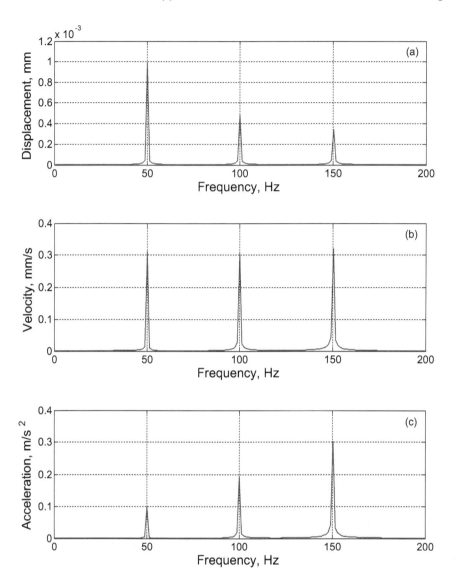

FIGURE 5.28 (a–c) Displacement, velocity, and acceleration spectra derived from one to another in frequency domain.

5.5 SHORT TIME FOURIER TRANSFORMATION (STFT)

The signal in Figure 5.22 is considered again to explain the STFT concept. It is also assumed that the time $t_l = 10$ s and the sampling frequency, $f_s = 1024$ Hz. If the segment size, $N = 2^{10} = 1024$ data and 50% overlap are used then there will be 19 segments (see Table 5.4) and hence 19 spectra (1 spectrum per segment). Instead of using these 19 spectra for a single averaged spectrum, if these 19 spectra were plotted into a

Signal Processing

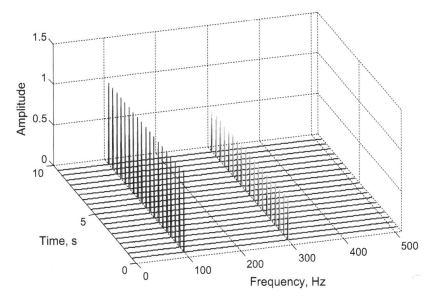

FIGURE 5.29 Waterfall plot of spectra for a time domain signal using STFT analysis.

3-D waterfall plot with x-y-z axes as frequency-time-amplitude respectively, then it is known as the STFT. Let us assume the 1st spectrum for the 1st segment is related to the meantime, $t_{s1} = \frac{0+1}{2} = 0.5$ s, similarly 2nd, 3rd, 4th, ..., 19th spectrum are assumed to associated with the mean time of each segment, $t_{s2} = \frac{0.5+1.5}{2} = 1.0$ s, $t_{s3} = \frac{1.0+2.0}{2} = 1.5$ s, $t_{s3} = \frac{1.5+2.5}{2} = 2.0$ s, ..., $t_{s19} = \frac{9.0+10.0}{2} = 9.5$ s respectively. The 3-D waterfall is shown in Figure 5.29. The advantage of this STFT process is to track how the frequencies and their amplitudes in a signal are changing with time. In this case, there are two frequency peaks at 100 Hz and 300 Hz with constant amplitude around 1 and 0.5, respectively, with time, hence there are no change in frequencies and their amplitudes in the signal from 0 to 10 s.

5.5.1 Example 5.4: An Experimental Rig

The experimental rig in Figure 4.26 and the same measured vibration acceleration data at a bearing are considered again. This time the STFT analysis is carried out. The computed STFT plot is shown in Figure 5.30. It is clear from the Figure 5.30 that both frequencies (40 Hz, 80 Hz, 120 Hz, 160 Hz, etc.) and their amplitudes are invariant with time.

5.5.2 Example 5.5: An Industrial Centrifugal Pump

The schematic of an industrial centrifugal pump is shown in Figure 5.31 (Balla et al., 2005). The complete assembly is supported by three steel blocks (bolted together) on either side. It is a vertical pump with an axial suction and a radial discharge as shown in the figure. It has six bladed impellers and total height from the

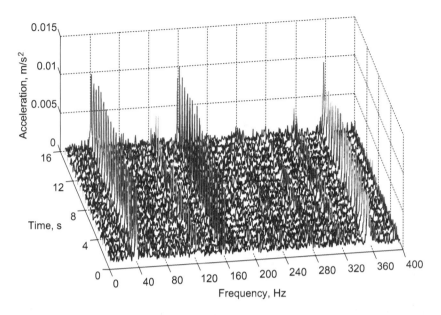

FIGURE 5.30 Measured vibration acceleration STFT plot at a bearing of the experimental rig in Figure 5.26(a).

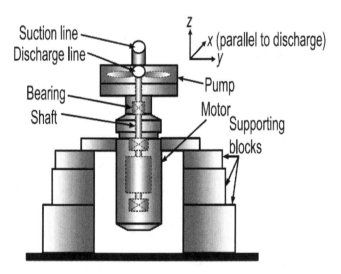

FIGURE 5.31 Schematic of an industrial centrifugal pump. (From Balla et al., *Adv. Vib. Eng.*, 4, 137–142, 2005.)

floor level to the pump top is more than two meters. The measured STFT plot on the pump casing is shown in Figure 5.32 when operating at 1500 RPM (25 Hz). The plot shows typically two start ups of the pump (Balla et al., 2005). This simply indicates the pump run-up and run-down were so quick that the STFT analysis has not pickup these transient operations. Both start-ups are showing operation at a

Signal Processing

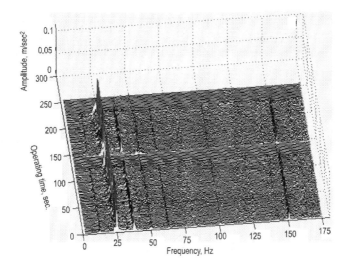

FIGURE 5.32 Measured STFT plot for the centrifugal pump in Figure 5.31.

constant 1500 RPM (25 Hz) and exactly same dynamic behavior, i.e., appearance of small peaks at 12.50 Hz, 37.5 Hz, 50 Hz, and 150 Hz (6 blades × 25 Hz) in addition to the frequency peak at 25 Hz (pump RPM).

Therefore the STFT analysis is useful to know the behavior of machines with time to understand whether the vibration behavior and/ or machine speed are changing or not. The STFT analysis may be good option if the machine speed is not constant during its operation to meet the industrial requirements.

5.6 CORRELATION BETWEEN TWO SIGNALS

The correction between two signals can be done both in time and frequency domains. However, the analysis in the frequency domain is only discussed here considering the use in the vibration monitoring.

5.6.1 Cross Power Spectrum

Equation (5.14) explains the concept of the averaged auto-power spectral density (PSD), $S_{yy}(f)$ for a time domain signal, $y(t)$. Similarly, the cross-power spectral density (CSD), $S_{xy}(f)$ at a frequency, f between two time domain signals, $x(t)$ and $y(t)$ is given mathematically as

$$\Rightarrow S_{xy}(f) = \frac{\sum_{i=1}^{n_s}(X_i(f)Y_i^*(f))}{n_s} \quad (5.16)$$

where $X_i(f)$ and $Y_i^*(f)$ are the FT and the complex conjugate of the FT at frequency, f for the i th segment of the signals, $x(t)$ and $y(t)$, respectively. "n_s" is the number of equal segments of the signals, $x(t)$ and $y(t)$, used for averaging.

5.6.2 TRANSFER FUNCTION (FREQUENCY RESPONSE FUNCTION)

The transfer function $x(t)/y(t)$ in the frequency domain i.e., $X(f)/Y(f)$ is known as the frequency response function (FRF). This is a useful tool in signal processing for comparing the relative amplitude and the phase of the signal, $x(t)$ with respect to the signal, $y(t)$ at any frequency, f. The averaged FRF is defined as

$$FRF(f) = \frac{S_{xy}(f)}{S_{yy}(f)} \tag{5.17}$$

$$\Rightarrow FRF(f) = \frac{\dfrac{\sum\limits_{i=1}^{n_s}(X_i(f)Y_i^*(f))}{n_s}}{\dfrac{\sum\limits_{i=1}^{n_s}(Y_i(f)Y_i^*(f))}{n_s}} \tag{5.18}$$

First, the averaged cross power spectral density (CSD) between the two signals, $x(t)$ and $y(t)$, and the averaged PSD of the signal, $y(t)$ are computed separately for the numerator and denominator of the FRF respectively. Then the averaged CSD is divided by the averaged PSD to compute the FRF. The computed FRF contains a complex quantity at each frequency which represents the amplitude ratio and phase of the signal $x(t)$ with respect to the signal, $y(t)$ at that frequency.

5.6.3 ORDINARY COHERENCE

The ordinary coherence between two vibration signals, $x(t)$ and $y(t)$, at a frequency is defined as

$$\gamma^2(f) = \frac{|S_{xy}(f)|^2}{S_{xx}(f)S_{yy}(f)} \tag{5.19}$$

$$\Rightarrow \gamma^2(f) = \frac{\dfrac{\left|\sum\limits_{i=1}^{n_s}(X_i(f)Y_i^*(f))\right|^2}{n_s}}{\left(\dfrac{\sum\limits_{i=1}^{n_s}(X_i(f)X_i^*(f))}{n_s}\right)\left(\dfrac{\sum\limits_{i=1}^{n_s}(Y_i(f)Y_i^*(f))}{n_s}\right)} \tag{5.20}$$

Signal Processing

The coherence between two signals indicates the degree to which two signals are linearly correlated at a given frequency. The scale of the coherence function is between 0 and 1. "0" at a frequency indicates no relation between the two signals at that frequency and "1" indicates prefect (100%) relation. The following can also be inferred about the coherence (Balla et al., 2006);

1. Coherence close to unity means the response is linearly correlated to the input excitation.
2. In general, measurement of structural responses and input are contaminated with noises (generated due to measuring instruments or/and structures) which may cause reduction in coherence.
3. Coherence reduces due to nonlinear relation between response and input.
4. Coherence reduces when the response is due to extraneous inputs.

5.6.4 EXAMPLE 5.6: TWO SIMULATED SIGNALS WITH NOISE

The following two signals are considered to illustrate the concept of the FRF and Coherence function.

$$x(t) = A_{x1} \sin(2\pi f_1 t + \theta_{x1}) + A_{x2} \sin(2\pi f_2 t + \theta_{x2}) + Noise \tag{5.21}$$

$$y(t) = A_{y1} \sin(2\pi f_1 t + \theta_{y1}) + A_{y2} \sin(2\pi f_2 t + \theta_{y2}) + Noise \tag{5.22}$$

The values of the parameters in Equations (5.21) and (5.22) are listed in Table 5.6. Both signals are assumed to be collected at the sampling frequency, $f_s = 1000$ Hz and the samples of the collected signals containing noises are shown in Figure 5.33. It is difficult to know the frequency contents and their respective amplitudes in both the signals based on the visual observation of signals in Figure 5.33 and the relation between these two signals. The FRF and coherence functions help to understand these relations. Hence, the FRF of the signal $x(t)$ with respect to $y(t)$ and their coherence are computed as per

TABLE 5.6

Parameters of the Signals, $x(t)$ and $y(t)$

Frequency, Hz	Amplitude	Phase, deg.
$f_1 = 100$	$A_{x1} = 10$	$\theta_{x1} = +90$
	$A_{y1} = 8$	$\theta_{y1} = +45$
$f_2 = 200$	$A_{x2} = 12$	$\theta_{x2} = -90$
	$A_{y2} = 6$	$\theta_{y2} = +45$

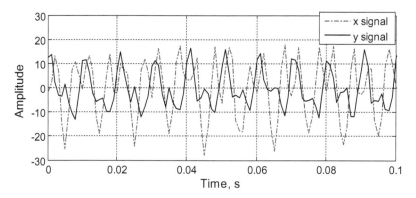

FIGURE 5.33 Typical sample of the signals, x and y, of Example 5.6.

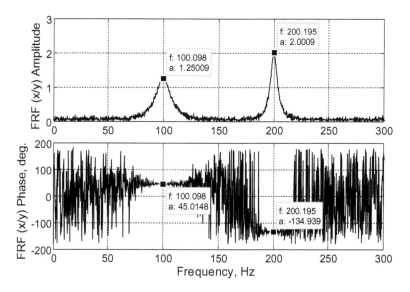

FIGURE 5.34 FRF amplitude and phase plot of Example 5.6.

Equation (5.18) and (5.20), respectively, which are shown in Figures 5.34 and 5.35. The following observations can be drawn from Figures 5.34 and 5.35:

1. FRF plot: Both signals just contain two frequencies—100 Hz and 200 Hz.
2. FRF amplitude: Amplitude ratios between the signals (x/y) at the frequencies 100 Hz and 200 Hz are 1.25 and 2, which are exactly the same as A_{x1}/A_{y1} and A_{x2}/A_{y2} respectively.
3. FRF phase: Phase angles of the signals, x with respect to y, at the frequencies 100 Hz and 200 Hz are $+45°$ and $-135°$ which are again exactly the same as $(\theta_{x1} - \theta_{y1})$ and $(\theta_{x2} - \theta_{y2})$, respectively.
4. Coherence: An amplitude of nearly equals to 1 at both 100 Hz and 200 Hz confirms that the both signals are well-correlated at these two frequencies.

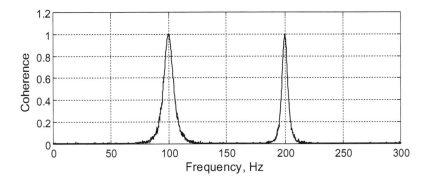

FIGURE 5.35 Coherence function plot of Example 5.6.

5.6.5 Example 5.7: Laboratory Experiments

The schematic of the laboratory experimental setup is shown in Figure 5.36. It consists of a clamped-clamped long tube made of steel and filled with a number of steel balls. A spring with a head is kept closed to the centre of the tube. The experiments and finding by Balla et al. (2006) are presented here. The tube was initially excited using a shaker connected to the pipe through a force sensor as shown in Figure 5.36. Both force and acceleration responses were measured simultaneously to estimate the FRF of the acceleration response, $x(t)$ to the applied force, $y(t)$ and the coherence between them.

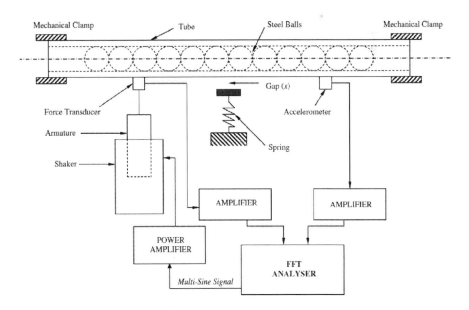

FIGURE 5.36 Schematic of the experimental setup of Example 5.7. (From Balla, C.B.N.S. et al., *J. Sound. Vib.*, 290, 519–523, 2006.)

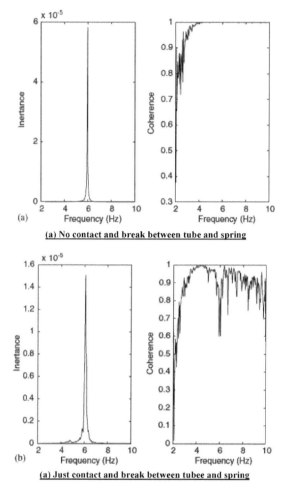

FIGURE 5.37 Measured FRF and coherence plots. (From Balla, C.B.N.S. et al., *J. Sound. Vib.*, 290, 519–523, 2006.)

Initially the excitation by the shaker was kept small to avoid interaction of the tube with the spring, and then shaker excitation was increased to allow just "contact and break" type of interaction between them.

The FRF and coherence plots for both cases are shown in Figure 5.37. The FRF plot is showing the natural frequency of the tube close to 6 Hz and the coherence amplitude equals to 1 when there was no interaction between tube and spring. This indicates system is completely linear as expected. However when the tube is interacting with spring (like contact and break), there is significant drop in the coherence but negligible change in the natural frequency. This interaction made the system non-linear and hence resulted in the reduction in the coherence.

Signal Processing

5.7 CONCEPT OF ENVELOPE ANALYSIS

Several types of signals are possible. Here only two types of signals, frequency modulation (FT) and amplitude modulation (AM), are shown in Figure 5.38. When the frequency of a signal is changing with time but no change in the amplitude then the signal is known as the FM signal. However, when there is no change in the frequency (time period of the signal) but change in the amplitude, then the signal is called the AM signal.

The AM occurs when there are many low frequencies with low energy levels get modulated with a signal of high frequency with high energy content. This high frequency with high energy is called the carrier frequency, f_c in the AM signal. Figure 5.38b is showing a typical AM signal where a low frequency with low amplitude, $f_m = 20$ Hz is get modulated with a carrier frequency, $f_c = 200$ Hz signal. The spectrum of this AM signal is shown in Figure 5.39. It is clearly showing 200 Hz but not 20 Hz. Instead 20 Hz is appearing as side band frequencies, $f_c - f_m = 180$ Hz and $f_c + f_m = 220$ Hz of the carrier frequency, $f_c = 200$ Hz.

This AM is a common phenomenon in the machine vibration signals. Often side bands are not clearly visible in the normal spectrum from the measured machine vibration signal and hence it makes difficult to do correct fault diagnosis typically for the anti-friction bearings. Since the carrier frequency and modulated frequencies may not be identified easily from the normal spectrum, it also makes difficult to perform the amplitude demodulation analysis at a selected carrier frequency for the signal. Therefore an alternate demodulation method is generally used in practice which is known as the "envelope" analysis.

The boundary of the modulated signal is shown by the dash line in Figure 5.40a. The boundary signal is called the envelope (demodulated) signal which removes the

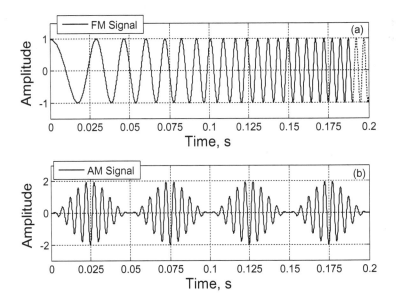

FIGURE 5.38 (a and b) Examples of the FM and AM signals.

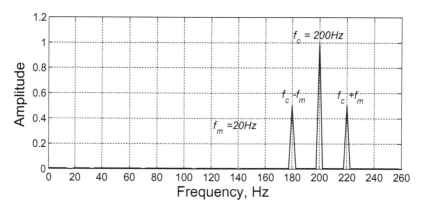

FIGURE 5.39 Spectrum of the AM signal shown in Figure 5.38(b).

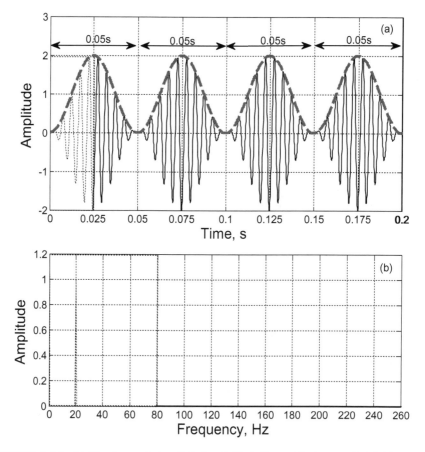

FIGURE 5.40 (a) Envelope of the AM signal (dash line with time period 0.05 s), (b) Spectrum of the envelope signal.

Signal Processing 111

affect of the carrier frequency. This envelope analysis for any signal can be done through the Hilbert transformation method (Bendat and Piersol, 1980). The time period of the envelope signal is also clearly marked in the Figure 5.40a. The period is equals to 0.05 s and hence the modulated frequency, f_m is equals to $1/0.05 = 20$ Hz. The spectrum of the envelope signal is also shown in Figure 5.40b, which shows the appearance of the modulated frequency of 20 Hz.

Hence the envelope signal and its spectrum can reveal the hidden frequency or frequencies due to the amplitude modulation in the signal.

5.8 SUMMARY

The chapter explains the step-by-step signal processing approaches for the vibration data in both time and frequency domains. A number of examples are also included to aid the understanding of the different signal processing methods.

REFERENCES

Balla, C.B.N.S., Jyoti K. Sinha, A.R. Rao, 2005. Importance of Proper Installation for Satisfactory Operation of Rotating Machines. *Advances in Vibration Engineering*, 4(2), pp. 137–142. (JK028).

Balla, C.B.N.S., K.K. Meher, Jyoti K. Sinha, A.R. Rao, 2006. Coherence Measurement for Early Contact Detection between Two Components. *Journal of Sound and Vibration*, 290(1), pp. 519–523.

Bendat, J.S., A.G. Piersol, 1980. *Engineering Applications of Correlation and Spectral Analysis*. Wiley, New York.

Elbhbah, K., Jyoti K. Sinha, W. Hahn, 2016. Investigation of High Piping Vibration, *Journal of Maintenance Engineering*, 1, pp. 7–15.

Jyoti K. Sinha, 2015. *Vibration Analysis, Instruments, and Signal Processing*. CRC Press/Taylor & Francis Group, January 2015. http://www.crcpress.com/product/isbn/9781482231441.

Oppenheim, A.V., R.W. Schafer, 1989. *Discrete-Time Signal Processing*, Prentice-Hall: Upper Saddle River, NJ.

6 Vibration Data Presentation Formats

6.1 INTRODUCTION

Different formats that are commonly used to present the analyzed vibration data typically for the machinery vibration-based condition monitoring are discussed in this chapter (Sinha, 2002, 2015). The analysis and presentation of the measured vibration data from a machine may be different for the machine running at a constant speed (i.e., normal machine operation) and for the machine transient operations such machine run-up and run-down operation (varying speeds). Hence these two conditions are discussed separately here.

6.2 NORMAL OPERATION CONDITION

During normal machine operation, the following data processing of the measurement vibration data and storage at predetermined intervals are important. The continuous monitoring even in case of permanently installed vibration-based monitoring system is not preferable. It is briefly discussed in Chapter 3 through the P-F curve. The continuous data analysis and their storage may not provide many benefits but to require more storage and data handling capabilities.

6.2.1 OVERALL VIBRATION AMPLITUDE

The overall vibration amplitude can be stored in terms of acceleration, velocity or displacement. These values are specified in three ways—root mean square (*RMS*), 0 to peak (*pk*) or peak-to-peak (*pk-pk*). For simple illustration, a displacement sine wave of 50 Hz, indicating the meaning of *RMS*, *pk*, and *pk-pk* values is shown in Figure 6.1a. The time waveform shown in Figure 6.1b consists of both sub-harmonic and higher harmonics of 50 Hz in the displacement signal. The *RMS* value of a sine wave is equal to 0.707 times the *pk* value, however, for any other kind of time waveform such as the one shown in Figure 6.1b, the *RMS* needs to be computed numerically (refer to Chapter 5).

6.2.2 VIBRATION SPECTRUM

The time domain vibration signals should be converted into the frequency domain. The plots of the spectrum should be stored for comparison and trending purpose from one measurement to another measurement to understand the machine vibration behavior. The vibration spectra of the time domain displacement signals in Figure 6.1 are shown in Figure 6.2. Since the machine speed is 3000 RPM, then the vibration amplitude at 50 Hz is referred to as 1×, 25 Hz as 0.5×, 100 Hz as 2×, and

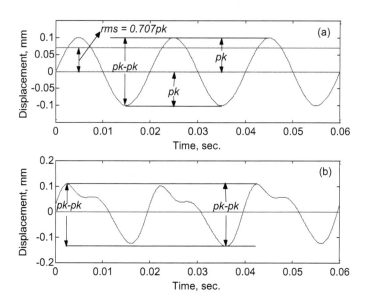

FIGURE 6.1 Typical displacement signals marked with peak, peak to peak, and RMS amplitude of vibration: (a) sine wave, (b) periodic wave.

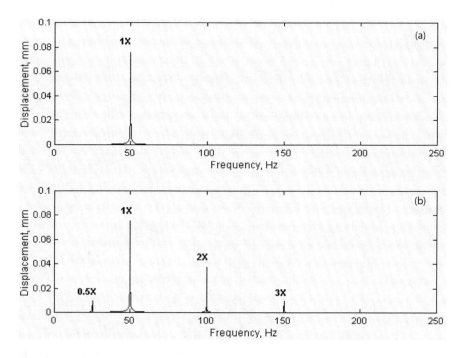

FIGURE 6.2 Spectra of the displacement signals in Figure 6.4(a) and (b).

Vibration Data Presentation Formats

so on as marked in Figure 6.2. Change in the amplitude at these frequencies should be monitored, as these changes may be related to some kind of fault development in the machine.

6.2.3 THE AMPLITUDE—PHASE VERSUS TIME PLOT

The 1× (or 2×, 3×,...) vibration amplitude and its phase may be plotted as a function of time for both the horizontal and vertical vibration at a bearing. A typical example of such a plot of 1× component for the vertical and horizontal shaft relative displacement with respect to the bearing housing and their phases with the time of machine operation is shown in Figure 6.3.

6.2.4 THE POLAR PLOT

Figure 6.3 can also be presented in a more compact way in polar coordinates as shown in Figure 6.4, which is called a polar plot. Thus only one polar plot for each bearing is sufficient to give the required information.

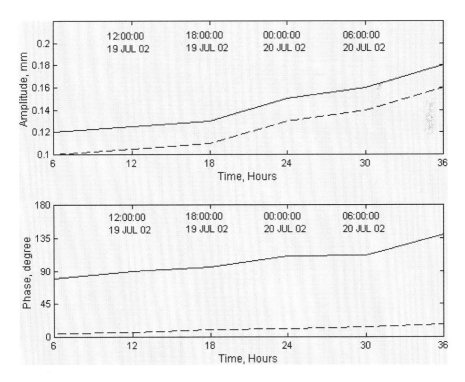

FIGURE 6.3 Variation of amplitude and phase of 1× component of shaft relative displacement with time. (___ Horizontal, _ _ _ Vertical.)

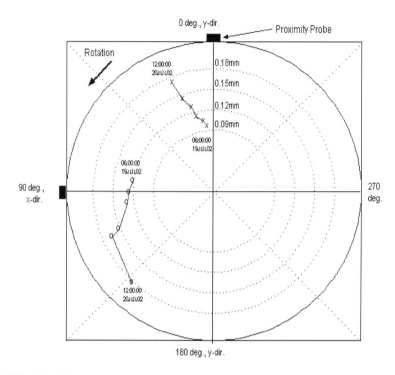

FIGURE 6.4 Polar plot representing the amplitude-phase vs. time data in Figure 6.3. (x, Horizontal; o, Vertical.)

6.2.5 THE ORBIT PLOT

The *x-y* proximity probes shown in Figure 4.15 of Chapter 4 are used to measure the relative shaft displacement to the stationary bearing pedestal. The proximity probes are generally used for the rotating machines where the rotors are supported through the fluid bearings. The signal from such a non-contacting proximity probe consists of the following two components.

1. A *dc signal* that is proportional to the average shaft position relative to the probe mounting. The *x-y* proximity probes give the shaft position in the bearing, and a change in shaft position during normal operation would indicate a load change or maybe bearing wear.
2. An *ac signal* corresponding to the shaft dynamic motion relative to the probe mounting. The *x-y* probe signals give the shaft centerline's path as the shaft vibrates. For illustration, Figure 6.5 shows an orbit plot obtained from the time waveform for the shaft relative displacement measured by *x-y* probes. A perfect circular orbit plot indicates that the supporting foundation stiffness is the same in the vertical and horizontal directions, and the rotor response is at 1× only. This plot is the forward orbit plot as the path of the rotor oscillation is in the direction of rotor rotation with respect to a keyphasor, otherwise it maybe a reverse orbit plot.

Vibration Data Presentation Formats

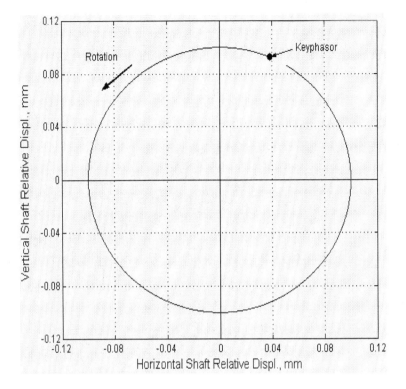

FIGURE 6.5 Orbit plot.

The orbit plot can be plotted either for the unfiltered time waveform signals or for the filtered component of 0.5×, 1×, 2×, etc. for each revolution of shaft. The shaft position and the orbit plots provide useful information about shaft malfunctions.

6.3 TRANSIENT OPERATION CONDITIONS

The start-up and run-down of a machine are considered as transient operation conditions. The measured data during this condition are very useful for confirming fault identification and its subsequent diagnosis. The following data analysis should be performed.

6.3.1 THE 3D WATERFALL PLOT OF SPECTRA

A 3-D waterfall plot is nothing but the STFT plot of the measured vibration signal as discussed in Chapter 5. A typical waterfall spectra plot for the measured vertical relative displacement of shaft for a machine during its startup typically from 2000 RPM to 3000 RPM is shown in Figure 6.6. The appearance or disappearance of any frequency components (multiples of 1×) or even a small change in the vibration amplitude at any frequency component can easily be tracked by this 3D waterfall plot of the spectra. In Figure 6.6, the time axis of the STFT plot is replaced by

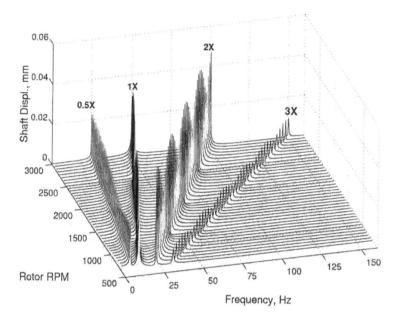

FIGURE 6.6 The waterfall plot of spectra during machine run-up.

the machine RPM. It is the user choice to use either time or speed in the waterfall plot if the relation between time and speed is known.

6.3.2 THE SHAFT CENTERLINE PLOT

Once again this analysis is useful for the rotating machines where the rotors are supported through the fluid bearings.

The plotting of dc signals of the *x-y* proximity probes at a bearing with machine rotating speed gives the shaft average centerline. This is useful for identifying changes in bearing load and bearing wear, as well as for calculating the average eccentricity ratio and the rotor position angle. The measured eccentricity helps in the understanding of fluid induced stability. An assumed shaft centerline plot during machine startup is shown in Figure 6.7. In Figure 6.7 the vertical and horizontal bearing clearance is 0.44 mm. At the start of the machine rotation the rotor was resting at bottom (180°) and then slowly moves up to 0.10 mm with an angle 225°. This indicates that the rotor is rotating in counterclockwise (ccw) direction and the eccentricity (ε) of journal center to bearing center is 0.12 mm.

6.3.3 THE ORBIT PLOT

The observation of orbit plots, both raw (unfiltered) and filtered 0.5×, 1×, 2×, etc. of *x-y* signals, with the change in machine speeds during transient operation are useful. This can help in predicting oil whirl, shaft rub, misalignment, crack, etc. (Sinha, 2015).

Vibration Data Presentation Formats

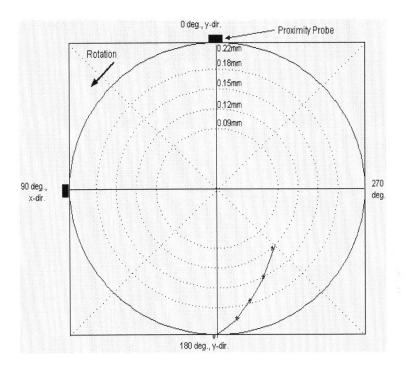

FIGURE 6.7 The shaft centerline plot.

6.3.4 THE BODE PLOT

The measured vibration data have to be order tracked to extract the 1× and higher harmonics with respect to the tacho signal. Order tracking gives the vibration amplitude-phase relationship of 1×, 2×, etc. with the change in the machine's rotating speeds. The plot of vibration amplitude and its phase with speeds is known as the Bode plot. A typical Bode plot of the 1× frequency component for a rotating machine, run-down from 2400 to 360 RPM, is shown in Figure 6.8.

The rotating speed is nothing but the source of vibration excitation for the machine due to unbalance. So if the machine speed is passing through the machine natural frequency or frequencies (critical speed or speeds) then resonance will occur and hence amplification in the vibration amplitude. In the plot shown in Figure 6.8, there are a total of four peaks—two in the vertical and two in the horizontal direction in the run-down frequency (speed) range. These may or may not be the natural frequencies of the machines. The phase angle of 1× vibration with respect to the tacho signal in both vertical and horizontal directions may be useful to identify the natural frequencies.

As per vibration theory discussed in Chapter 2 (Figure 2.10), the phase change should be 180° from before and after the natural frequency. Here all four peaks (2 in vertical and 2 in horizontal directions) clearing showing 180° changes in the phase angle at each peak. Hence these four frequencies are the natural frequencies which are also commonly known as the machine critical speeds.

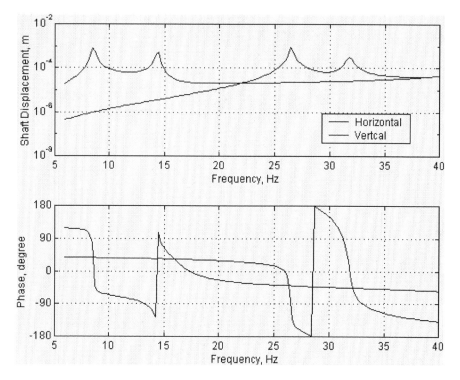

FIGURE 6.8 The Bode plot of 1× shaft displacement for a machine run-down.

6.4 SUMMARY

Commonly used data formats to present the analyzed machinery vibration data are discussed in this chapter.

REFERENCES

Sinha, J.K. 2002. Health Monitoring Techniques for Rotating Machinery, PhD Thesis, University of Wales Swansea (Swansea University), Swansea, UK, October 2002.

Sinha, J.K. 2015. *Vibration Analysis, Instruments, and Signal Processing.* CRC Press/Taylor & Francis Group, January 2015. http://www.crcpress.com/product/isbn/9781482231441.

7 Vibration Monitoring, Trending Analysis and Fault Detection

7.1 INTRODUCTION

Figure 7.1 shows an abstract representation of any rotating machines. Any rotating machines used in industries may consists of three major components—rotor, bearings and their foundation—that may include all supports, auxiliary components, casing, piping, etc. The rotors are of different types such as motor rotor, turbine rotor, pump rotor, compressor rotor, generator rotor, etc. depending upon the rotating machines. In many machines, two or more rotors are connected through the gearboxes.

As explained in Chapter 2, the rotor unbalance force is responsible for the vibration in any machine. This force generates only the speed synchronous vibration which is represented by $1\times$ (1 time machine speed) in the measured vibration spectrum. However, if there is any fault(s) in a machine, it may generate vibration at frequencies related to sub and higher harmonics of the machine speed ($0.5\times$, $1\times$, $15.5\times$, $2\times$, etc.), depending upon the fault types.

The measurement locations, directions of measurement, and measurement interval are also discussed in Chapter 3. The abstract of the regular stage-2 vibration-based condition monitoring (VCM) procedure of Chapter 3 is shown here again in Figure 7.2. The selection of the instrumentation and sampling frequency are discussed in Chapter 11 based on the knowledge from this chapter.

The first step is to measure the machine vibration and then estimate the overall vibration velocity (i.e., overall velocity, V_{RMS}) at each measurement location in the machine. The estimated V_{RMS} at each measurement location then needs to be compared with the recommended vibration limit as per the ISO code or the limits provided by the machine manufacturers. The ISO code on machine vibration is briefly discussed in Chapter 3. It is also important to understand why ISO has recommended the vibration limits in velocity, but most of machine vibration measurements are generally carried out by accelerometer sensors.

To explain the reason for this clearly, the measured vibration data at a bearing for a machine are considered here for three conditions, namely Condition-1 Healthy, Condition-2 Small Defect and Condition-3 Large Defect size. The measured vibration spectrum plots are shown in Figures 7.3 through 7.5 for Conditions-1 to 3, respectively. Each figure consists of the displacement, velocity and acceleration spectrum plots for the one-to-one comparison. The methods in Chapter 5 are used to

122 Industrial Approaches in Vibration-Based Condition Monitoring

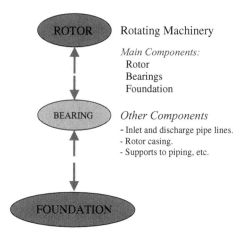

FIGURE 7.1 An abstract representation of a rotating machinery.

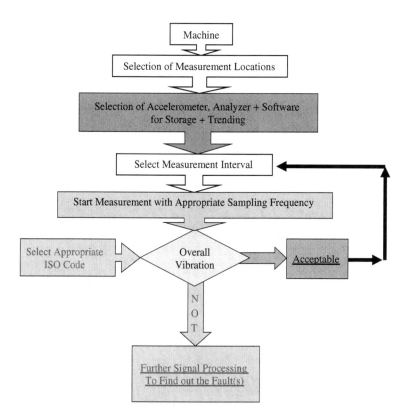

FIGURE 7.2 Stage 2: An abstract of the VCM procedure during the machine useful life.

Vibration Monitoring, Trending Analysis and Fault Detection

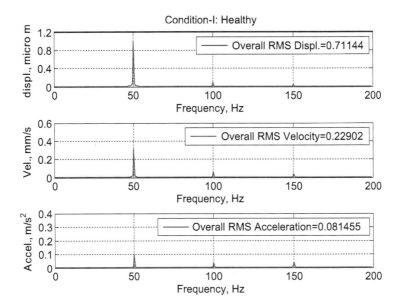

FIGURE 7.3 Comparison of the vibration displacement, velocity and acceleration spectra for a healthy machine.

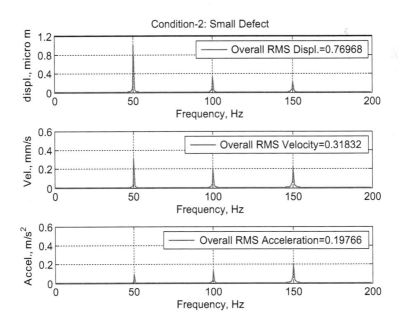

FIGURE 7.4 Comparison of the vibration displacement, velocity and acceleration spectra for a machine with a small size defect.

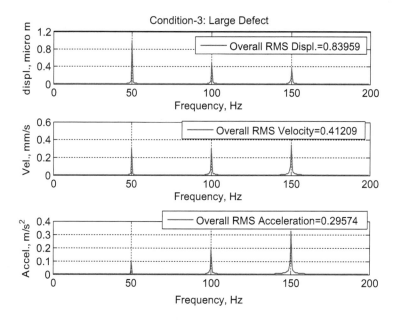

FIGURE 7.5 Comparison of the vibration displacement, velocity and acceleration spectra for a machine with a large size defect.

TABLE 7.1
Overall RMS Vibration Values

Condition	Displacement, μm	Velocity, mm/s	Accelerometer, m/s²
1- Healthy	0.711	0.229	0.081
2- Small Defect size	0.769	0.318	0.197
3- Large Defect size	0.839	0.412	0.295

convert the measured acceleration spectrum to the velocity and displacement spectra. The overall RMS vibration values for these three cases are also listed in Table 7.1 for the direct comparison.

It is obvious from Table 7.1 that the change in overall RMS displacement from Conditions 1 to 3 is not significant but change in the RMS acceleration is significant. However, the change in the RMS velocity is nearly proportional to the machine conditions. It is also evident from the displacement spectrum plots in Figures 7.3 through 7.5 that there are no significant changes in vibration amplitudes at 1× (50 Hz), 2× (100 Hz) and 3× (150 Hz) from Conditions 1 to 3, respectively. It is because the vibration displacement decreases with the increase in vibration frequency. Therefore, the change in acceleration amplitudes is likely to be significantly high at higher frequencies. It is because the acceleration amplitude is equals to ω^2 times the displacement amplitude at the frequency (ω). However, the vibration velocity amplitude at a

Vibration Monitoring, Trending Analysis and Fault Detection

frequency is the product of ω (frequency) and the displacement, hence the velocity is better parameter to estimate the machine condition. If the overall vibration V_{RMS} is higher than acceptable or allowable limit, then it is better to do further vibration data analysis as per the steps shown in Figure 7.2 to understand the machine vibration behavior and identify if any fault (or defect) is present in the machine or not.

The vibration-based detection of different faults in machines is discussed in this chapter using the concept of vibration theory, instrumentations and measurements, data analysis, etc. that are already discussed in the earlier chapters.

7.2 TYPES OF FAULTS

There can be many faults that develop or keep on developing in a rotating machine during its operation. However, the failure of machines in the last few decades has always identified the probable cause for the failure in the post-failure study. Many failures have been related to the system design and many others to human error—deviation in assembly and inaccuracy in manufacturing, as well as variation in the machine operation conditions, etc. So, there have been consistent efforts, for many years, on design modifications and the development of fault identification tools to mitigate possibility of machine failures. The machine performance shows that there has been significant improvement in the design process as current machines rarely fail due to design problems. However, other kinds of problems due to human error and deviation in operating conditions need to be avoided. There are many well recognized faults resulting from other kinds of errors than the design, which are listed here.

1. Rotor faults—mass unbalance, shaft bent or bow, misalignment, crack, shaft rub, etc.
2. Bearing (anti-friction and fluid bearings) faults
3. Gearbox faults
4. Motor faults, etc.

7.3 ROTOR FAULTS DETECTION

The identification features and procedures for each well-known rotor fault commonly used in industries are outlined here in a simple manner. The vibration features of different rotor faults, mechanical looseness and rotor blade (impeller) defects are also summarized in Table 7.2.

7.3.1 MASS UNBALANCE

In practice, rotors are never perfectly balanced because of manufacturing errors such as porosity in casting, non-uniform density of material, manufacturing tolerances, and loss or gain of material (e.g., scale deposit on the blades, material loss due to erosion, etc.) during operation. Obviously, these factors often result in the high centrifugal force in the rotor and hence high vibration in the machine. This centrifugal force is nothing but the rotor unbalance force. It is explained in Chapter 2.

TABLE 7.2
Features of The Rotor Related Faults

Fault Type	Spectrum Features	Features of Orbit Plots
Mass unbalance	1× Amplitude ↑ trend, possible change in 1× phase	Circular or elliptical orbit depending on support stiffnesses in the measured directions
Shaft bend or bow	Features similar to the unbalance. 1× Amplitude ↕ trend depending upon the plane of bending. Axial 1× amplitude ↑ trend.	
Misalignment	Presence of 1×, 2×, 3×, … (their amplitudes may be invariant over a period of time)	Banana-type (small misalignment) to figure 8 shape if excessive misalignment for the orbit plots consist of 1× and 2× components
Shaft crack	Presence of 1×, 2×, 3×, … (their amplitudes are likely to increase and significant change in their phases due to propagation of the crack size over a period of time).	Refer to Figure 7.7
Shaft rub	Presence of ×/4 or ×/3 or ×/2 and its higher harmonic components.	Better to have orbit plots consist of ×/4 or ×/3 or ×/2 component only to understand the rub phenomenon.
Mechanical looseness	Generally, start with the presence of 1×, 2×, 3×, … and then leads to 0.5×, 1×, 1.5×, 2×, 2.5×… with increase in the looseness.	
Impeller/blade damage	Presence of 1BFP, 2BPF (the increase in their amplitudes with the propagation of damage size). BPF = Machine Speed × No of Impeller Blades.	

This unbalance force generally generates a dominant 1× component of the rotor vibration during normal operation. The polar plots of the 1× component for both horizontal and vertical directions at all bearings may show increase in amplitude with or without any significant change in phase with time.

It is good to trend the vibration velocity at 1× component. At any time of machine operation if the overall vibration velocity RMS, V_{RMS} exceeds or close to the allowable limit only due to 1× vibration then the rotor balancing is recommended to reduce the machine vibration to the acceptable limit. The accepted practice for the rotor unbalance estimation is known as the *influence coefficient method*, which is explained in Section 7.10.

7.3.2 SHAFT BOW OR BEND

Bent or bowed shafts may be caused in several ways, for example due to creep, thermal distortion or a previous large unbalance force. The forcing caused by a bend is similar, although slightly different, to that caused by mass unbalance. The shaft

Vibration Monitoring, Trending Analysis and Fault Detection

vibration spectra are generally dominated by the 1× component. The best way to detect the presence of the shaft bow is to trend the axial vibration overall and 1× component.

It is always better to rectify the shaft bend and then perform the rebalancing before returning back the machine to operation. Alternatively, if the bend is small and the correction is likely to be expensive, then just perform the rotor balancing at the operating speed. This is likely to reduce the machine vibration at the operating speed, but it may not balance the rotor at all frequencies in the machine run-down speed range.

7.3.3 Misalignment

The misalignment occurs at the coupling location of two shafts, or relatively small off-sets in the bearings position during machine assembly process after maintenance activity. In practice the shaft misalignment may be of three types—parallel, angular or coupled misalignment (Figure 7.6). It is generally accepted that a significant 2× component together with 1× is indication of the shaft misalignment. However, it is often observed that the misalignment generates 1×, 2×, 3×,...components. The trending of the amplitudes of these components may not change significantly with machine operation time. It is because the misalignment is expected to be remain unchanged over a period of time of machine operation. The orbit plot of the shaft relative displacement vibration signals (measured by the proximity probes) consisting of 1× and 2× components may be close to a banana-type shape or the shape of a figure 8 if misalignment is excessive.

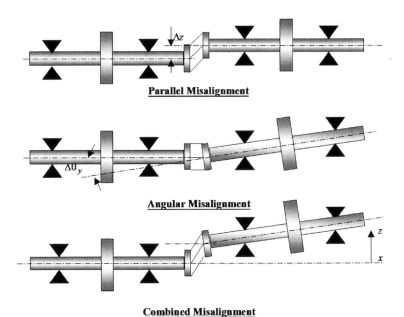

FIGURE 7.6 Possible types of rotor misalignment.

7.3.4 SHAFT CRACK

If a transverse crack develops in a shaft, the stiffness of the shaft will vary from high-to-low-to-high in a complete rotation of the shaft caused by breathing (opening and closing) of the crack due to the rotor self-weight. This behavior of the crack shaft also generates a 2× component similar to shaft misalignment. But unlike the misaligned shaft, both the amplitude and phase of the 1× and 2× components change with the time of machine operation due to the propagation of the transverse crack. This can be observed in either the polar plots or the amplitude-phase versus time plots. The orbit plot consisting of 1× and 2× components may change from a single loop to double loops like a figure 8. However, during transient operation of the machine, the shaft vibration may be very high when the machine passes through nearly half of the machine critical speed. At this particular moment, the shape of the orbit plot will change from a figure 8 to a loop containing a small loop inside indicating a significant change in the phase and amplitude of shaft vibration as shown in Figure 7.7.

7.3.5 SHAFT RUB

Rotating machine rubs occur when a rotor contacts the machine stator. Rubs are generally classified as a secondary malfunction as they may typically be caused by primary malfunctions; like a poorly balanced rotor, turbine blade failure, defective bearings and/or seals, rotor misalignment, bowed shaft either mechanical or thermal, deformed casing, etc. Rub is often of the hit-and-bounce type behavior of a rotor. The behavior of the system during this period is highly non-linear and may also be chaotic.

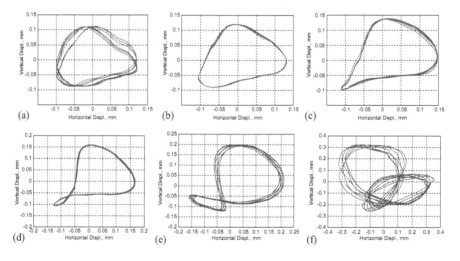

FIGURE 7.7 Orbit plots at different rotor speed during the small rig run-up. (a) 300 RPM, (b) 450 RPM, (c) 600 RPM, (d) 650 RPM, (e) 700 RPM and (f) 750 RPM.

Vibration Monitoring, Trending Analysis and Fault Detection 129

Since it is a complex phenomenon but the general observation is that the rub generates subharmonic ×/2, ×/3 components in the vibration spectrum. A typical example of such a phenomenon is discussed in Section 7.7.

7.4 OTHER MACHINE FAULT DETECTION

7.4.1 MECHANICAL LOOSENESS

Bolts, joints and the bearing assembly may loosen over a period of machine operation. This looseness initially generates 1×, 2×, 3×, etc. components but since looseness can propagates with machine operation and hence, sub-harmonics (×/2 or ×/3) components may start appearing together with 1×, 2×, 3×, etc. components during normal operation of machines. Hence trending of these data is important to identify the defect correctly.

7.4.2 BLADE PASSING FREQUENCY (BPF)

The BPF is calculated using the rotor speed times number of blades. The vibration measurements on the bearing pedestals/housing are likely to contain the blade vibration for the smaller machines like pumps, fans, blowers, compressors, etc. Hence the monitoring of the vibration amplitudes of the BPF, and its higher harmonics are important. The increase in amplitudes at these frequencies with time may indicate the propagation of the looseness in the blade assembly and/or damage in the blades.

7.4.3 BLADE VIBRATION AND BLADE HEALTH MONITORING (BHM)

The use of the BPF may not be useful for the blades in machines like turbo-generator (TG) sets. Hence the BHM for such machines are not straightforward. The shaft Torsional Vibration and the Blade Tip Timing (BTT) methods are found to be useful for this requirement. The BTT method is an in-direct approach to measure and monitor the deflection of each blade in the bladed discs during machine operation. It is an intrusive method that uses a number of non-contact sensors circumferentially mounted on the machine casing just above the blades' tip to measure each blade arrival time. The measured arrival times acquired by the different sensors are then used to estimate each blade vibration displacement.

7.4.4 ELECTRIC MOTOR DEFECTS

In an electric motor, the three possible defects, namely (a) general electric problem (GEP), (b) stator winding defect (SWD) and (c) rotor defect (broken bar) can occur during its operation. Hence the monitoring and detection of these faults are essential. If the line frequency and motor speed (in Hz) are given by f and f_r respectively, then these three defects can be detected by the following features in the vibration spectrum (Table 7.3).

TABLE 7.3
Features of The Motor Defects

Faults	Features in Vibration Spectrum	Further Details
General electric problem	nf and nf_r	where $n = 1, 2, 3,...$
Stator winding defect	$f, 2f, 4f,...$	Even harmonics of line frequency
Rotor defect (broken bar)	$f \pm 2nsf, 2f \pm 2nsf,...$	where s = slip and $n = 1, 2, 3,...$

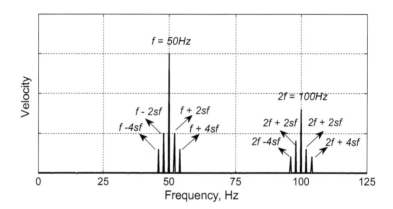

FIGURE 7.8 Typical vibration spectrum for a motor with broken rotor bars.

A typical vibration spectrum for the motor rotor defect (broken bars) is shown in Figure 7.8 for a motor where $f = f_r = 50$ Hz. Appearing of more sidebands at $f, 2f$, etc., and increase in their amplitudes with machine operation time may indicates the propagation of the motor rotor defect size.

7.5 GEARBOX FAULT DETECTION

Gearbox is one the commonly used mechanical device in power transmission. The gear ratio (GR) is one of the technical specifications of any gearbox. This is nothing but the ratio of speed of the output shaft to input shaft of the gearbox. The gearbox can be of single stage or multi-stages. A pair of wheels meshes together through a number of teeth on each wheel shown in Figure 7.9 is known as a single stage gearbox. Here the one gear wheel is connected to the input (drive) shaft and other to the output (driven) shaft. The speeds of the input and output shafts are related through the Equation 7.1.

$$N_1 T_1 = N_2 T_2 \qquad (7.1)$$

where (N_1, N_2) are the rotating speeds of the gear wheel 1 (input) and 2 (output), and (T_1, T_2) are the number of teeth on wheel 1 and 2, respectively. Hence the gear ratio will be given by

Vibration Monitoring, Trending Analysis and Fault Detection

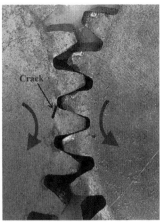

FIGURE 7.9 Photograph of a gear pair and its zoomed view showing possible crack at a tooth root during gear mesh.

$$\text{GR} = \frac{N_2}{N_1} \tag{7.2}$$

To aid the easy understanding of the gearbox vibration response, a healthy single stage gearbox as shown in Figure 7.10 is considered where the gear wheel-1 is called as the pinion wheel ($N_1 = 100$ RPS, $T_1 = 10$) and the gear wheel-2 as the gear wheel ($N_2 = 10$ RPS, $T_2 = 100$). The vibration measurement is carried out on the bearing B1 as shown in Figure 7.10. The measured vibration responses are also shown in Figure 7.10 for one complete rotation of the pinion shaft. The vibration responses shown in the figure is consisting of the pinion shaft response plus the gear meshing. The pure sinusoidal response of the pinion shaft indicates the shaft is subjected to only unbalance condition and ten complete cycles also indicate the meshing of ten pinion teeth with the gear wheel teeth during one full rotation of the pinion shaft. The resultant vibration response is also shown in Figure 7.10. Therefore, the vibration spectrum is likely to have 100 Hz (related to the pinion speed) and 1000 Hz (related to tooth meshing). The vibration spectrum is shown in Figure 7.12a also confirms the presence of 100 Hz and 1000 Hz. This 1000 Hz is known as the gear mesh frequency (GMF) or tooth mesh frequency (TMF). It can be calculated for each pair of gear wheels as

$$f_{GMF} = N_1 T_1 = N_2 T_2 \tag{7.3}$$

Now consider the same gearbox again but with a crack at the root of a tooth (first tooth out of ten teeth) in the pinion wheel. If there is a crack at the tooth root (as shown in Figure 7.9) then the tooth will have more deflection due to crack opening when it meshes with other gear wheel tooth. Hence the vibration deflection for the cracked tooth is likely to be more compared to remaining nine teeth deflection. This is typically shown in Figure 7.11. The nature of this vibration of tooth meshing in case of crack is likely to have amplitude modulation response as shown in

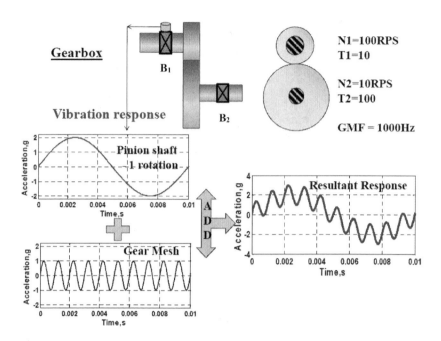

FIGURE 7.10 Typical vibration of a healthy gearbox.

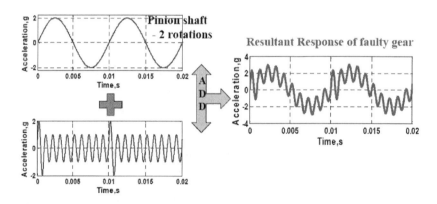

FIGURE 7.11 Typical vibration of a gearbox with a cracked tooth in the pinion.

Figure 7.11 where there is no change in the time period of each tooth meshing, but amplitude became higher for the cracked tooth.

The spectrum of this vibration response is also shown in Figure 7.12b. The spectrum is showing 100 Hz (1× of the pinion rotating speed), 1000 Hz (the GMF) plus modulation frequency of 100 Hz as the sidebands of 1000 Hz. Therefore, the presence of 100 Hz as the sidebands clearly indicates the defect or crack in the pinion wheel tooth/teeth.

If the crack in a tooth or teeth is related to the gear wheel of speed 10 Hz, then the sidebands will appear at space of 10 Hz around the GMF 1000 Hz (such as …980 Hz, 990 Hz, 1000 Hz, 1010 Hz, 1020 Hz,…).

Vibration Monitoring, Trending Analysis and Fault Detection 133

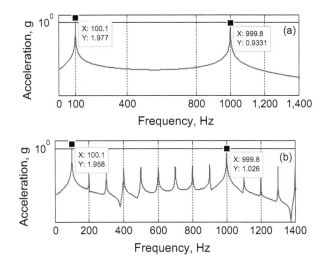

FIGURE 7.12 Spectrum plots of a gearbox, (a) healthy condition, (b) faulty condition (a cracked tooth in the pinion).

The crack in a tooth or teeth is going to propagate with time of operation due to breathing nature (opening and closing) of crack and hence the sideband frequencies amplitudes will increase with time. Typical vibration spectrum plots for different conditions of a gearbox are shown in Figure 7.13, and their summaries are listed in Table 7.4.

Figure 7.14 is also showing a typical vibration spectrum when the gear wheels are not assembled exactly same as earlier after maintenance work. The appearance of the hunting frequency and/or the assembly phase frequency together with its harmonics simply indicates the same teeth of both gear wheels are not meshing again after re-assembly. It is always good to benchmark the dis-assembly and re-assembly process to avoid such vibration and other malfunction issues for any machines. The hunting frequency (f_H) and the assembly phase frequency (f_{APF}) are calculated as

$$f_{APF} = \frac{f_{GMF}}{n_{pf}} \tag{7.4}$$

$$f_H = \frac{f_{GMF} n_{pf}}{T_1 T_2} \tag{7.5}$$

where n_{pf} is the common prime factors of number of teeth in both gear wheels such as $T_1 = n_{pf} n_1$ and $T_2 = n_{pf} n_2$.

It is often suggested to do amplitude demodulation to known the exact modulated frequency or frequencies. The envelope analysis discussed in Section 5.7 of Chapter 5 can be used for this purpose. However, the envelope analysis should not be applied to the measured raw vibration signal blindly. The steps for the envelope analysis are shown in Figure 7.15.

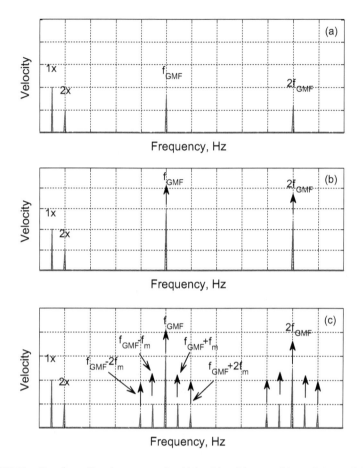

FIGURE 7.13 Gearbox vibration spectra for (a) healthy, (b) wear, (c) crack in tooth/teeth.

TABLE 7.4
Features of Different Gearbox Conditions

Conditions	Features in Vibration Spectrum	Further Details
Health gearbox	$f_{GMF}, 2f_{GMF}, 3f_{GMF},\ldots$	Possibility of small peaks at $2f_{GMF}, 3f_{GMF},\ldots$ in addition to f_{GMF} peak
Tooth/teeth wear, assembly looseness, etc.	$f_{GMF}, 2f_{GMF}, 3f_{GMF},\ldots$	Amplitude values are likely to increase at frequencies $f_{GMF}, 2f_{GMF}, 3f_{GMF},\ldots$
Tooth/teeth crack	$f_{RMS} \pm nf_m,$ $2f_{RMS} \pm nf_m,\ldots$	where f_m is either input or output gear wheel speed in Hz.
In correct assembly	$f_{GMF}, 2f_{GMF}, 3f_{GMF},\ldots$ plus $f_H, f_{APF}, 2f_{APF},\ldots$	where f_H and f_{APF} are the hunting and assembly phase frequencies respectively. This means the same teeth are not meshing again after re-assembly.

Vibration Monitoring, Trending Analysis and Fault Detection

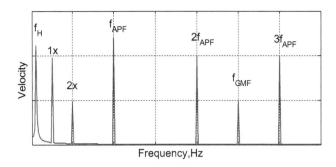

FIGURE 7.14 Typical spectrum of a gearbox showing the hunting frequency and assembly phase frequency with its harmonics if not assembled correctly.

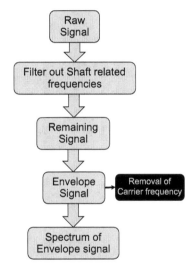

FIGURE 7.15 Steps involved in an envelope analysis.

Step 1: Remove the rotor related frequencies (1×, 2×, 3×, etc.) by the high pass filter of the measured raw data. The remaining data (i.e., filtered signal) will have the only frequencies that are related to GMF frequencies and modulated frequencies. For this gearbox, the carrier frequency will be the GMF of 1000 Hz and the modulated frequency, f_m will be 100 Hz in case of the amplitude modulated signal.

Step 2: Apply envelope analysis to the remaining signal to obtain the envelope signal.

Step 3: Estimate spectrum of the envelope signal to find the frequency or frequencies of the modulated signal or signals with the carrier frequency.

This process is typically shown in Figure 7.16 through the gearbox with a cracked tooth in the pinion wheel. The dashed line is showing the envelope time domain signal and its spectrum clearly shows the presence of the modulated signal of 100 Hz.

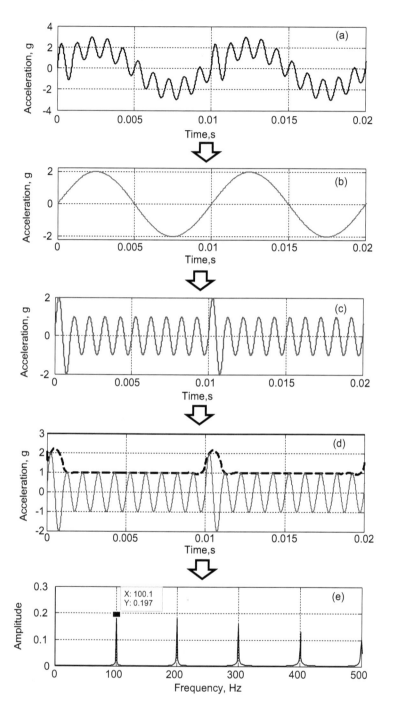

FIGURE 7.16 A typical example for the envelope analysis for a gearbox, (a) raw signal, (b) removal of rotor signal, (c) remaining signal after removal of rotor signal, (d) envelope signal (dashed line), and (e) spectrum of the envelope signal.

7.6 ANTI-FRICTION BEARING FAULT DETECTION

An anti-friction bearing can be a ball or roller bearing. A simple schematic of a ball bearing is shown in Figure 7.17. It consists of an inner race, an outer race, a cage and a number of balls. The defect can occur in any of these components during the machine operation. Any defect in anti-friction bearing causes an impulsive loading due to metal-to-metal contact generated by the presence of defect in the bearing. This metal-to-metal impact generates excitation up to a high frequency band in order of a few kHz during each revolution of the shaft. Hence this impulsive force per revolution is going to generate decay kind of vibration response per revolution. This decay vibration response is nothing but most likely to be the resonance response of the bearing assembly and housing. Hence the accelerometer mounted on the bearing housing is going to pick the vibration of the shaft plus the decay vibration response of the bearing housing and assembly resonance. Typically, such vibration responses are shown in Figure 7.18 considering the rotor is subjected to the rotor unbalance only.

A typical vibration acceleration spectrum for a ball or roller bearing is shown in Figure 7.19. The presence of the frequency components 1×, 2×, ... indicates the vibration related to the shaft; however, the broadband peak in high frequency region is likely to be the bearing assembly and housing resonance. The appearance of this banded high frequency peak may indicate the presence of bearing defect or lack of lubrication. The lack of lubrication also generates metal-to-metal rub and hence excite the bearing resonance. The resonance frequency is generally in the order of few kHz. The value obviously depends on the bearing size, assembly and housing but it may be up to 5 or 6 kHz. It is always good to do the *in situ* hammer test on the bearing housing assembly to know the resonance frequency. A typical example is shown in Figure 7.20. Impulsive forces by the hammer are used to excite the bearing assembly and then measure the vibration responses by the accelerometer. The measured spectrum is clearly showing a hump around 4.2 kHz, which is nothing but the

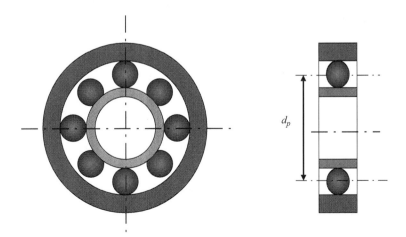

FIGURE 7.17 A simple schematic of a ball bearing.

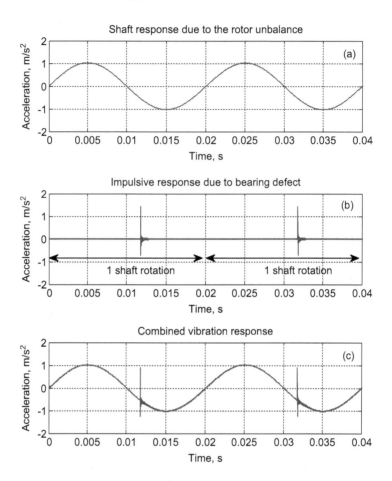

FIGURE 7.18 (a) Shaft response, (b) impulsive response, (c) measured vibration response (addition of (a) and (b)) for a typical faulty anti-friction bearing.

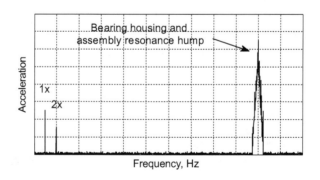

FIGURE 7.19 A typical measured acceleration spectrum on anti-friction bearing housing with a bearing defect.

Vibration Monitoring, Trending Analysis and Fault Detection 139

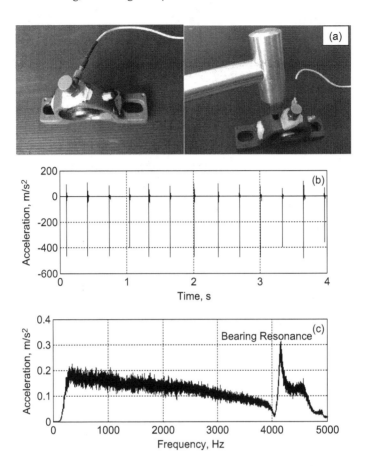

FIGURE 7.20 A typical example of the resonance response of a bearing assemblies and housing due to impact forces, (a) test setup, (b) measured acceleration response, (c) its spectrum.

bearing assembly resonance frequency. It is always good to perform *in situ* hammer tests on each bearing assembly in the machine to know their resonance frequencies.

As discussed earlier (Section 7.1), it is good to have the overall vibration velocity V_{RMS} and velocity spectrum to carry out the vibration-based monitoring and fault diagnosis. However, it is always better to use vibration acceleration signal for the diagnosis of anti-friction bearing defects. The reason is explained through a simple example of the measured vibration acceleration signal shown in Figure 7.21. The spectrum plots (both linear and log y-scale) are also shown in Figure 7.21. It is obvious from the spectrum that the vibration acceleration signal contains peaks at the frequencies 50 Hz, 100 Hz and 2000 Hz (2 kHz). This vibration acceleration signal is also converted into the velocity and displacement signals, which are shown in Figures 7.22 and 7.23, respectively, for comparison. It is obvious from the spectrum plots in Figures 7.22 and 7.23 that the frequency peak at 2 kHz is not clearly visible definitely in the displacement spectrum, but a very small visible peak in the velocity

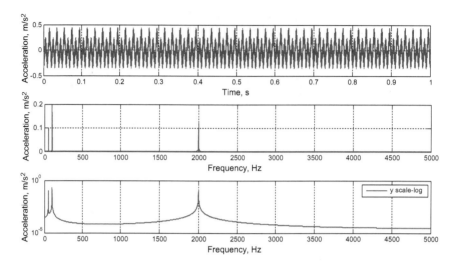

FIGURE 7.21 A typical measured vibration acceleration time domain signal and its spectrum (y-scale—linear and log).

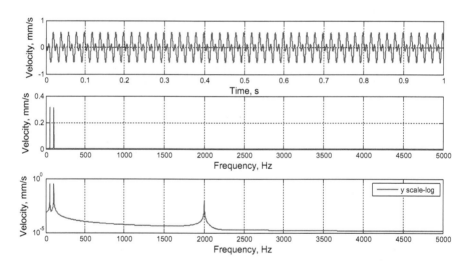

FIGURE 7.22 Vibration velocity signal derived from the acceleration signal in Figure 7.21 and its spectrum (y-scale—linear and log).

spectrum. It is because the acceleration, velocity and displacement amplitudes at a frequency are related by equation, $a(f) = (2\pi f)v(f) = (2\pi f)^2 d(f)$. So, when the frequency is high then the displacement is going to be small.

Therefore the high-frequency impulsive responses may not be dominant in the vibration displacement and velocity signals; hence it is better to use acceleration signal for the anti-friction bearing defect diagnosis. Such impulsive acceleration responses can be quantified and picked by the following non-dimensional parameters.

Vibration Monitoring, Trending Analysis and Fault Detection 141

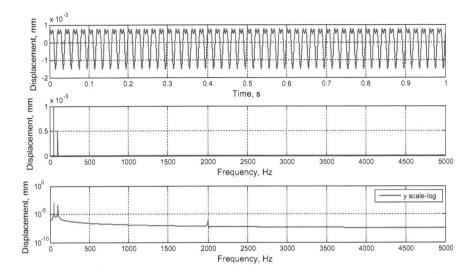

FIGURE 7.23 Vibration displacement signal derived from the acceleration signal in Figure 7.21 and its spectrum (y-scale—linear and log).

7.6.1 CREST FACTOR (CF)

It is the ratio of the peak to the RMS acceleration (Section 5.1) of the bearing vibration If a machine response is purely sinusoidal at its RPM, the value of CF will be 1.414. The CF value can increase to a high value if impulsive responses are present in the vibration signal due to the bearing defect.

7.6.2 KURTOSIS (Ku)

It is another non-dimensional parameter to identify the impactness in any signal. It is defined as the fourth order statistical moment (Section 5.1) for any signal. The *Kurtosis (Ku)* value for the sine wave is 1.5, whereas the value is 3.0 for the random waveform with Gaussian amplitude distribution. The impulsive signal from the bearing is likely to have the kurtosis value more than 3.

The trending of the CF or Kurtosis can detect the bearing defect. A typical trend of the CF or Kurtosis in case of anti-friction bearing defect is shown in Figure 7.24. It is good to establish the baseline or alarm limit based day-to-day measurements when the bearing is in good condition. If any defect is initiated in the bearing, it will generate impulsive loading and hence the CF/Ku may cross the alarm limit. Once the defect is initiated, it is likely to be propagating (growing) further with machine operating time and hence increases in the impulsive force and therefore both CF and Ku will show increasing trend. But once the defect size is further widened, the impulsive force is likely to reduce with time though the defect size is still growing and hence both CF and Ku will show decreasing trend after a certain defect size. When the defect size is wide spread and it may not generate any impulsive force and then lower values of both CF and Ku.

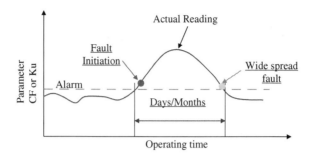

FIGURE 7.24 Trending of non-dimensional Crest Factor (CF) or Kurtosis (Ku) for bearing defect detection.

Both CF and Ku are tools to detect impulsiveness in the signal. This quality of the CF and Ku are utilized for the bearing diagnosis; therefore, it is suggested to carefully observe the CF or Ku trend and take the appropriate maintenance action before both CF and Ku fail to detect defect in the bearing.

7.6.3 Envelope Analysis

The CF and kurtosis are fully dependent on the impulsive nature of the signal. If the impulsive response is not due to the bearing defect then both CF and Ku are not capable to distinguish whether it is due to bearing defect or not. However, the envelope analysis of the measured bearing acceleration vibration signal is found be better option in identifying the bearing defect even at very early stage itself.

Let d_b, d_i, and d_o are the diameter of the balls, the inner race (OD), and the outer race (ID) respectively, n is the number of balls for a ball bearing shown in Figure 7.17. The characteristic frequency of each part is given by:

$$\text{Frequency related to the ball defect, } f_b = \frac{d_p}{d_b} f_r \left(1 - \left(\frac{d_b}{d_p}\right)^2 \cos\beta^2 \right) \text{ Hz} \quad (7.6)$$

$$\text{Frequency related to defect in the inner race, } f_i = \frac{n}{2} f_r \left(1 + \frac{d_b}{d_p} \cos\beta \right) \text{ Hz} \quad (7.7)$$

$$\text{Frequency related to defect in the outer race, } f_o = \frac{n}{2} f_r \left(1 - \frac{d_b}{d_p} \cos\beta \right) \text{ Hz} \quad (7.8)$$

$$\text{Frequency related to defect in the cage, } f_{cage} = \frac{f_o}{n} \text{ Hz} \quad (7.9)$$

where d_p is the pitch circle diameter, β is the contact angle for the balls, and f_r is the relative speed (Hz) between the inner race and the outer race, which is generally equal to the RPM of the shaft. The same formula is also valid for roller type bearings.

Vibration Monitoring, Trending Analysis and Fault Detection

If there is no defect in the bearing then none of these bearing characteristics frequencies will be excited. But if there is defect in the outer race or inner race or ball defect or cage defect then the bearing vibration will have one or all of these frequencies depending upon the defect in one component or all four components simultaneously. However, the spectrum of the bearing acceleration vibration may not able to show peaks at the bearing defect frequencies and hence the detection process is not straightforward.

It is because the defect bearing frequency or frequencies get amplitude modulated with the bearing housing and assembly resonance frequency and hence not visible in the normal spectrum. Since the bearing resonance is likely to appear as the broadband peak (like hump), it is difficult to see the sidebands of the modulated frequency or frequencies. Hence the demodulation is required around the bearing resonance frequency to detect the presence of any defective frequency/frequencies. Here again the steps listed in Figure 7.15 and discussed in Section 7.5 should be followed.

7.7 EXPERIMENTAL EXAMPLES

7.7.1 EXAMPLE 7.1—ROLLER BEARING DEFECT

The photograph of an experimental rig is shown in Figure 7.25. The rig shaft is connected to the motor through a flexible coupling and supported through a roller bearing. The photograph of the defect on the inner race is also shown in Figure 7.25. The bearing details are listed Table 7.5. The data from vibration

FIGURE 7.25 Photographs of the experimental rig and defect in the inner race of the roller bearing.

TABLE 7.5

Bearing Details

Characteristic Property	Value
Pitch circle diameter (d_p)	62.71 mm
Diameter of roller (d_b)	11.91 mm
Contact angle of the roller (β)	0
Number of rollers (n)	10
Inner (bore) diameter	40 mm

measurements carried out by Okogie et al. (2017) are used here. The measured vibration acceleration spectrum up to 260 Hz is shown in Figure 7.26a when rig was rotating at 1200 RPM (20 Hz). The spectrum is clearly showing harmonics of 20 Hz. The inner-race defect frequency of 118.99 (around 120 Hz) is estimated from Equation 7.7. The peak at this 120 Hz is also seen in the spectrum (Figure 7.26a). Hence it is difficult to conclude whether the peak at 120 Hz (6×) is present due to the rotor fault or bearing fault.

The envelope analysis may provide the answer to this observation. The measured vibration data is initially high pass filtered at 2 kHz to remove all frequency components up to 2 kHz. Figure 7.26b shows the spectrum of the filtered signal (remaining signal) where there is not frequency peaks up to 2 kHz. There is a cluster around 5–7 kHz, but it is not obvious whether they are related to the bearing resonance; hence the envelope analysis is carried out on this filtered (remaining) signal. The spectrum of the envelope signal is also shown in Figure 7.26c, which clearly reveals the peaks at the inner race and out race frequencies of 120 Hz and 80 Hz, respectively. Hence this confirms the presence of the inner race defect. It is good to examine the outer race as well.

7.7.2 EXAMPLE 7.2—ROTOR FAULTS

Part of the study by Sinha and Elbhbah, 2013 is briefly discussed here. An experimental rig shown in Figure 5.26 of Chapter 5 is again considered here. Typical measured vibration spectra at bearing B1 are shown in Figure 7.27. The shaft misalignment was introduced at coupling between the rig and motor shafts. A Perspex block as shown in Figure 7.28 is used to simulate shaft rub. The spectrum plot for the health condition shows peaks at 1× and a few higher harmonics for the healthy condition as expected. It is because achieving ideal healthy machine condition in reality is difficult. The healthy spectrum only indicates the possibility of a small shaft misalignment and residual unbalance; however, as expected the misalignment case shows higher harmonics of 1× component and the components of the sub-harmonics 0.5× (hump near around 17.0 Hz) for the shaft rub condition.

Vibration Monitoring, Trending Analysis and Fault Detection

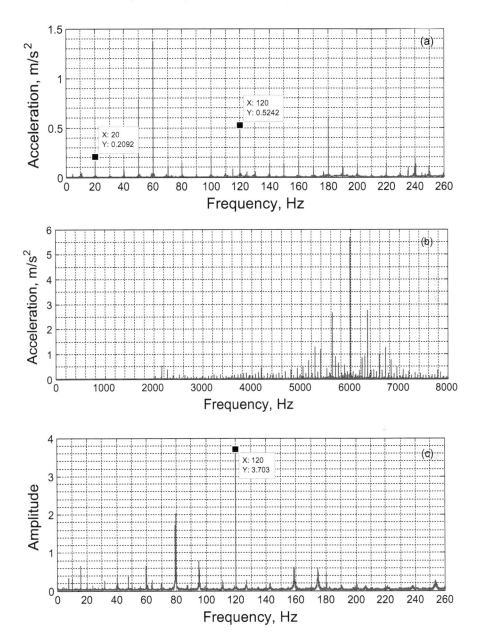

FIGURE 7.26 (a) Measured vibration acceleration spectrum up to 260 Hz, (b) spectrum plot up to 8000 Hz but high pass filtered up to 2000 Hz, (c) spectrum of envelope signal of (b).

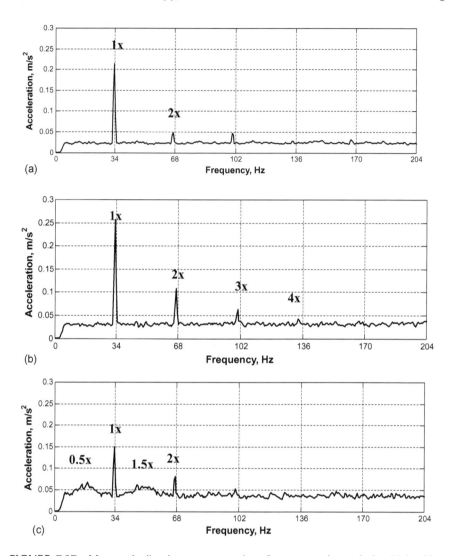

FIGURE 7.27 Measured vibration spectrum plots for an experimental rig, (a) healthy, (b) shaft misalignment, (c) shaft rub conditions.

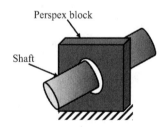

FIGURE 7.28 Experimental simulation of the shaft rubs.

7.8 INDUSTRIAL EXAMPLES

7.8.1 EXAMPLE 7.3—FAN WITH UNBALANCE PROBLEM

This is a typical industrial case study of the primary air fan (PA-Fan) used in a coaled fired power plant (Sinha et al. 2015). The high vibration in the machine is observed after maintenance which resulted into the motor bearing failure and hence the stoppage of power production.

Fan is used for supplying primary air to the mills, then conveying pulverized fuel into the boiler unit. The PA-Fan is having ten blades, supported by two journal bearings and driven by a 1.2 MW induction electric motor. The motor is mounted on two journal bearings. Figure 7.29 shows the photograph of this fan-motor unit and its schematic.

The overall vibration velocity RMS found to be much higher after maintenance. It was significantly higher than the unacceptable limit of 7.1 mm/s (RMS) as per the ISO 10816-3 code at the motor drive end bearing (B3). This high vibration finally leads to the bearing failure of drive end motor bearing over the period of fan operation. The failed bearing photographs are also shown in Figure 7.29.

The measured vibration spectrum at the bearing B3 before failure in Figure 7.30 clearly shows the dominant peak at 1×. This indicates the possibility of rotor unbalance and the unbalance mass may be close to the bearing B3. This highlights the issue with machine re-assembly after maintenance activity. Hence the coupling was

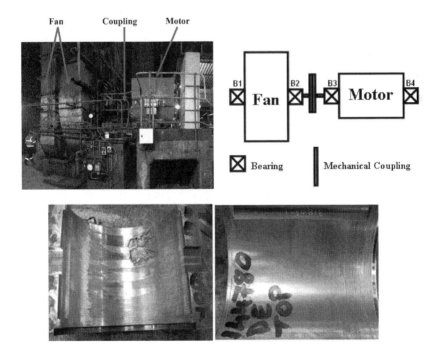

FIGURE 7.29 Photograph and schematic diagram of primary air (PA) fan, and photograph of failed drive-end motor bearing.

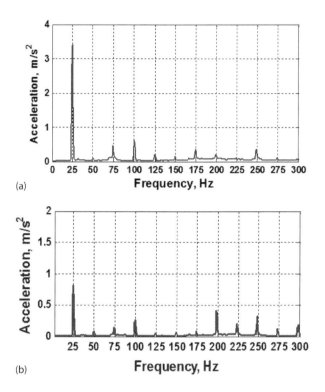

FIGURE 7.30 Typical measured vibration spectra at the motor drive end bearing (B3) before (a) and after (b) corrections.

step-by-step rotated during the coupling process (assembly) to observe the impact on the rotor balancing. It is observed to have significant impact on the balancing and vibration. With this exercise an optimum coupling location gave satisfactory result. Overall vibration reading came down to nearly 3 mm/s (RMS velocity), which is within the satisfactory limit as per the ISO 10816 code. The measured acceleration vibration spectrum after this unbalance correction by rotating the coupling position in Figure 7.30 also shows significantly reduced vibration at 1×.

Once again it is important to benchmark the machine before dismantling for maintenance and repair so that the machine can re-assemble correctly.

7.8.2 Example 7.4—Gearbox Fault

This is another industrial study by Asnaashari et al. (2016). A typical fan-gearbox-motor (FGM) unit photograph with schematic diagram is shown in Figure 7.31. This fan is used in the water-cooling tower of a power plant for the forced cooling of the condenser water. The input shaft of the gearbox is connected to the motor and the gearbox output shaft to the fan shaft. The gearbox used here is two-stage speed reduction—from the motor to intermediate shaft speed and then intermediate shaft speed to the fan shaft speed.

Vibration Monitoring, Trending Analysis and Fault Detection 149

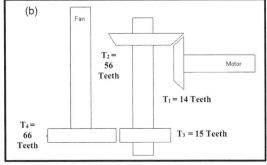

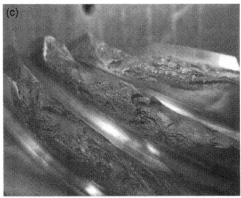

FIGURE 7.31 (a) Photograph of a motor-gearbox-fan unit, (b) schematic of the unit, and (c) failed gear teeth photograph.

The running speeds of the gearbox, fan and motor and their related frequencies together with the gear mesh frequencies (GMFs) are listed in Table 7.6. The measured acceleration spectrum is shown in Figure 7.32a. The spectrum shows the presence of the high GMF (HGMF) at around 348 Hz and its higher harmonics, which indicates the wear in the teeth related to the motor and intermediate shafts gear mesh. The closer look around high GMF 348 Hz in Figure 7.32b shows the presence of sidebands of the cluster of frequency peaks at the equal interval of 6.2 Hz. These sidebands indicate the presence/initiation of fault in both the motor

TABLE 7.6
Running Speeds and Gear Mesh Frequencies

		Running Speeds/Frequencies	
Motor speed	N_1		1485 RPM, 24.75 Hz
Intermediate shaft speed	$N_2 = \dfrac{N_1 T_1}{T_2} = \dfrac{24.75 \times 14}{56} = N_3$		371.2 RPM, 6.18 Hz
Fan speed	$N_4 = \dfrac{N_3 T_3}{T_4} = \dfrac{6.18 \times 15}{66}$		84.38 RPM, 1.4 Hz
Low GMF (between fan and intermediate gear shaft)	$f_{LGMF} = N_3 T_3 = N_4 T_4$		92.4 Hz
High GMF (between motor and intermediate gear shaft)	$f_{HGMF} = N_1 T_1 = N_2 T_2$		346.5 Hz

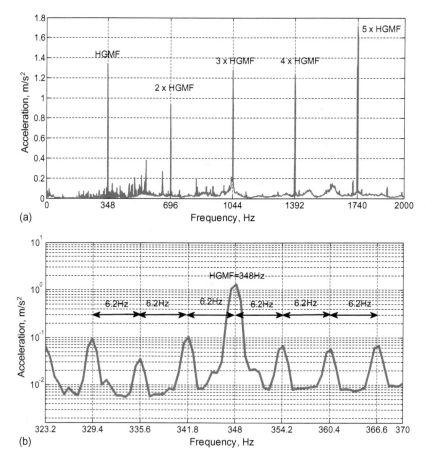

FIGURE 7.32 (a) Measured vibration acceleration spectrum on the gear box, (b) zoomed view at the HGMF = 348 Hz clearly showing modulation with the frequency of 6.2 Hz.

and intermediate shafts gear teeth. This is because the intermediate shaft speed is 317.2 RPM (6.18 Hz) and the motor speed is 1485 RPM (24.75 Hz, which is equal to 4 × 6.18 Hz). Note that the increase in the amplitude of sideband frequency peaks with machine operating time generally indicates the fault propagation with time. Inspection also reveals the failure of teeth in both motor and intermediate shafts gear pair.

7.9 MACHINES HAVING FLUID BEARINGS

There are many machines where the rotors are supported through the fluid bearings. The fault diagnosis methods discussed earlier are also applicable to this machine except the bearing diagnosis. The possible problems in the fluid bearings and their detection processes are discussed in this section.

A fluid-induced instability is likely problem in any fluid bearings. This fluid induced instability is commonly referred to as *oil whip*, which is results from *oil whirl*. Both typically occur in fluid bearings. This is actually a fluid-induced self-excited phenomenon leading to lateral shaft vibration. Oil whip occurs when the shaft rotates in the bearing and the fluid surrounding the journal (see Figure 7.33) also rotates at some circumferential speed $(\lambda\Omega)$ relative to the shaft speed (Ω), where λ is known as the fluid average velocity ratio. The ratio λ is generally a non-linear function of the fluid bearing radial stiffness, which depends on the shaft eccentricity ratio. It has been observed experimentally and analytically that the fluid bearing radial stiffness increases with the journal radial deflection and the velocity ratio λ decreases from maximum to zero i.e., journal is in direct contact with the wall of bearing. The fluid resonance frequency of the fluid in the bearing during shaft rotation is normally slightly less than half of the shaft rotating speed, say $0.48\,\Omega$. If during machine operation or start-up the fluid circumferential speed comes close to the fluid resonance frequency and instability of the shaft occurs that is known as *oil whirl*. This is generally identified by the presence of a $\lambda\times$ (usually 0.45–0.48× or less) component in the spectrum and

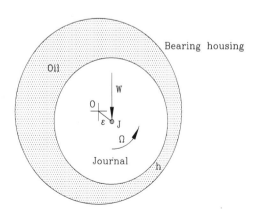

FIGURE 7.33 Schematic of a simple journal bearing.

the forward circular orbit plot for the filtered $\lambda\times$ component irrespective of the asymmetric stiffness in the vertical and horizontal direction. If the rotor system natural frequency is equal to the fluid resonance then the system would be completely unstable. The orbit plot may then be of the order of the bearing clearance. This phenomenon is known as *oil whip*. *Oil whip* can occur at the first lateral mode as well as at higher modes of the rotor. *Oil whip* generally occurs during machine start-up. It initiates with *oil whirl* and then gets locked to the *oil whip* when the shaft speed passes through a critical speed. A typical example during a machine start-up is shown in Figure 7.34.

Rotor systems in many machines have many fluid bearings and if a fluid-induced instability is observed during machine operation it important to identify the bearing that is causing such instability. The phase relation at the frequency of instability ($\lambda\times$) of the shaft relative displacement between the measurements at all the bearings can be used to identify the bearing. For the bearing which is suspected as the source of instability, its shaft relative displacement at the $\lambda\times$ frequency component will have a phase lag with respect to the shaft relative displacement at other bearings.

There are different fluid bearings commercially available to reduce this fluid instability problem. They are of the lemon-shaped bearings with grooves and multi-lobes, rough journal surfaces, and bearings with mobile parts (such as tilting pad bearings), option of fluid anti-swirl injections in bearings instead of just plain cylindrical bearings, to avoid the increase in the fluid circumferential velocity (Muszynska and Bently, 1996).

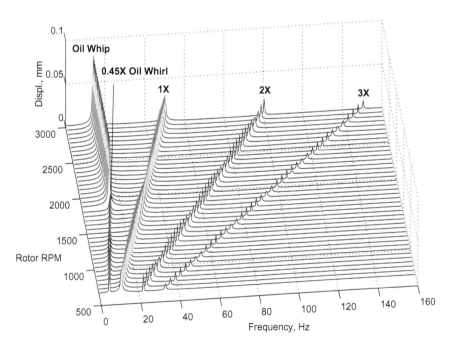

FIGURE 7.34 A typical case of oil whirl—oil whip during a machine run-up.

Vibration Monitoring, Trending Analysis and Fault Detection

7.10 FIELD ROTOR BALANCING

As discussed earlier in Section 7.3.1, the rotors are never perfectly balanced due to several reasons. The presence of dominant 1× component of the shaft vibration during machine normal operation is indicative of the rotor unbalance. This is generated due to synchronous rotor unbalance force. Hence the machine with high 1× vibration may need balancing to reduce the vibration within acceptable limit. The accepted practice for unbalance estimation is known as the *influence coefficient method or sensitivity method*, which is explained here through a single plane balancing.

7.10.1 SINGLE PLANE BALANCING—GRAPHICAL APPROACH

To aid understanding, a simple example of a short length rotor with bladed disc typically found in pumps, fans, compressors, etc. is considered here. A typical simplified form of a representative rotor with a balance disc and a bearing is shown in Figure 7.35. The balancing exercise is discussed in the following steps.

Step 1—Response measurement: It is assumed that the shaft is rotating anticlockwise at 3000 RPM. The amplitude of vibration at 1× is 52 μm at 50 Hz (3000 RPM) at an angle of 30° from the reference signal of the tacho sensor. The time waveform of the measured vibration signal of the 1× component (after band pass filtering around 50 Hz) is shown in Figure 7.36 together with the tacho signal. The point-to-point on the tacho signal represents one complete rotation of the shaft (i.e., 360°). The time for one complete shaft rotation is equals to 0.02 s. The point of the tacho signal represents the 0° angle. Hence the angle of 1× vibration can be estimated from the plots in Figure 7.36. It is the angle between the maximum 1× response to the tacho reference point (0°) as shown in Figure 7.36. A stroboscope can also be used to measure this angle.

Step 2—Definition of unbalance plane and amplitude: Let's assume that the unbalance response at 1× (**Ov**) of 52 μm@ 30° at an operating speed of 50 Hz (3000 RPM) is observed due to the unbalance (centrifugal) force, $m_{ub}r\omega^2$ acting on the balancing disc, where m_{ub} is the rotor unbalance mass

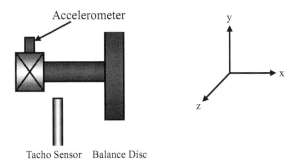

FIGURE 7.35 Schematic of a rotor with a balance disc together with measuring instruments.

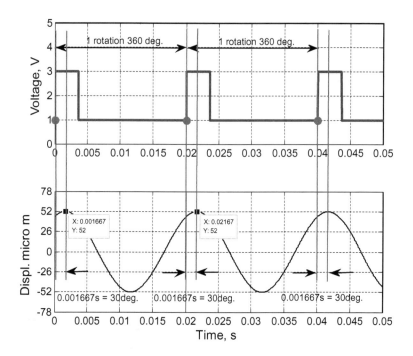

FIGURE 7.36 Measured vibration displacement signal (1×) together with tacho signal before balancing.

on the balance plane, r is the radius from the shaft center for the unbalance mass and $\omega = 2\pi f$ is the shaft angular speed. It is essential to estimate the unbalance mass (m_{ub}) and its location (both radial, r and angular position, θ_{ub}) to balance the rotor. It is also assumed that both the radius, $r = 50$ mm and the shaft speed of 3000 RPM (50 Hz) are constant during the balancing exercise, hence the only unknown are the unbalance mass, m_{ub} and angular position, θ_{ub}.

Step 3—Trial run: The influence of the known unbalance on the bearing vibration response is important to know if one has to estimate the unknown rotor unbalance. This requires trial run with known trial unbalance mass. Let us assume a trial unbalance mass, $m_t = 1$ gram at the radius, $r = 50$ mm, and angle, $\theta_t = 90°$ is added to the balance disc. The measured vibration at 1× now becomes equal to 90.21 μm @ 60° at the machine speed of 3000 RPM. The 1× vibration of 90.21 μm @ 60° is named as "**Ov+Tv**" because this is the resultant response due to the original unbalance, $m_{ub}r\omega^2$ plus added unbalance, $m_t r\omega^2$.

Step 4—Polar plot analysis: Once the required measurements are done, then do the analysis on the polar plot to estimate the unbalance and angle. Both **Ov** and **Ov+Tv** are marked onto the polar plot in Figure 7.37, where the radius of the circle indicates the vibration amplitude together with the phase angle. Since the shaft rotation is anti-clockwise so the angle on the polar plot

Vibration Monitoring, Trending Analysis and Fault Detection

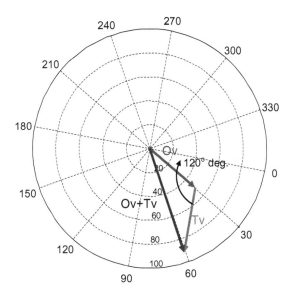

FIGURE 7.37 Polar plot for the rotor unbalance estimation.

must be clockwise. The arrow **Tv** (= 52 µm@ 120° with the arrow **Ov**) can then be estimated from the polar plot. The arrow **Tv** indicates the change in machine vibration due to the trial mass, hence the influence on bearing vibration due to the trial mass is called Sensitivity (S) which is given by

$$S = \frac{|Tv|}{m_t} = \frac{|Ov|}{m_{ub}} \qquad (7.10)$$

Thus, $\qquad m_{ub} = \dfrac{|Ov|}{|Tv|} m_t = \dfrac{52}{52} \times 1 = 1 \text{ gram}$

Step 5—Rotor balancing: Now the final step in the balancing needs to be carried out on the machine balancing disc. The arrow **Tv** makes an angle of 120° with the arrow **Ov** in the clockwise in the polar plot (see Figure 7.37). Hence, the estimated mass of 1 gram has to be added to the balance disc at the radius of $r = 50$ mm from trial mass location to 120° clockwise to balance the rotor. This is shown in Figure 7.38. The trial mass must be removed during the balancing process. Now the machine is supposed to be balanced.

7.10.2 Single Plane Balancing—Mathematical Approach

The same balancing example is now demonstrated through the mathematical approach.

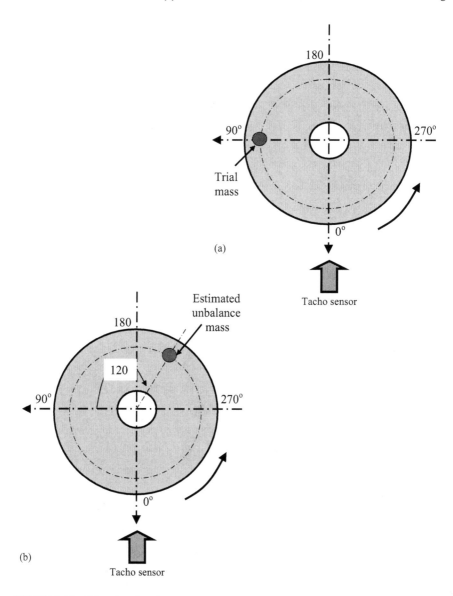

FIGURE 7.38 Disc showing location of trail mass and estimated unbalance mass (a) trial mass of 1 gram added at 90° (b) estimated unbalance mass of 1 gram added at 120° from trial mass location.

Step 1—Response measurement: Exactly same as the earlier, $\mathbf{Ov} = 52$ μm@ $30° = 4.5108 \times 10^{-5} + 2.6037 \times 10^{-5} j$ m.

Step 2—Definition of unbalance plane and amplitude: The unbalance force is written as $m_{ub} r \omega^2 e^{j\theta_{ub}}$, where ω is the angular speed of the rotor. If this unbalance is known, the mass (m_{ub}) can then be added at the disc radius (r), at an angle $(180 + \theta_{ub})$ to balance the rotor. Hence, the unbalance can be

Vibration Monitoring, Trending Analysis and Fault Detection

expressed as the complex quantity, $e = m_{ub}e^{j\theta_{ub}} = e_r + je_i$, where the radius, r and the machine speed are constant parameters during the balancing process.

Step 3—Trial run: Same as earlier, $\mathbf{Ov+Tv} = 90.21$ μm @ $60° = 4.5113 \times 10^{-5} + 7.8121 \times 10^{-5} j$ m due to addition of trial mass at the radius, $r = 50$ mm, $e_t = 1$ gram @ 90 degree $= m_t e^{j\theta_t} = 0 + j$ gram.

Step 4—Construction of sensitivity and unbalance estimation: The sensitivity matrix (Equation 7.10) is computed as

$$S = \frac{\mathbf{Ov}}{e} = \frac{(\mathbf{Ov+Tv}) - \mathbf{Ov}}{e_t} = 5.2124e\text{-}05 - 7.1679e\text{-}08j \text{ m/gram}$$

Therefore, the unbalance can be estimated as

$$e = S^{-1} \times \mathbf{Ov} = 0.8633 + 0.5000\ j = \text{nearly } 1 \text{ gram @ } 30 \text{ degree}$$

Step 5—Rotor balancing: The estimated unbalance is 1 gram at an angle of $30°$ from the tacho reference. Hence, this unbalance mass is to be added at an angle of $180 + \theta_{ub}$ degree $= 210°$ from the tacho reference to balance the rotor.

The balancing is exactly the same as the graphical method as shown in Figure 7.38. However, the advantage of this mathematical approach is that it can include a number of measurements on the rotor in comparison to the graphical approach. This can further be extended to multi-planes balancing.

7.11 SUMMARY

The chapter provides the comprehensive summary for the faults' detection processes in rotating machines. A few experimental examples and industrial examples are also discussed to demonstrate the detection of different faults. The step-by-step approach for a single plane rotor balancing is also demonstrated.

REFERENCES

Asnaashari, E., Elbhbah, K. Hahn, W. Jyoti K. Sinha, 2016. Avoiding Unplanned Shutdown by Simple Vibration-based Condition Monitoring, *Maintenance & Asset Management Journal* 16(4), 36–37.

Okogie, S., Jyoti K. Sinha, K. Elbhbah, A. Ojetunde, 2017. Vibration-Based Diagnosis of Faults in the Split-type Roller Bearing, *Journal of Maintenance Engineering* 2, 494–508.

Sinha, J.K., K. Elbhbah, 2013. A Future Possibility of Vibration based Condition Monitoring, *Mechanical Systems and Signal Processing* 34(1–2), 231–240.

Sinha, J.K., Keri Elbhbah, E. Asnaashari, I. Andrew, W. Hahn, 2015. Re-Assembly Impacting Machine Performance: A Case Study, *Maintenance & Asset Management Journal* 30(6), 47–49.

8 Experimental Modal Analysis

8.1 EXPERIMENTAL PROCEDURE

The natural frequencies and modeshapes together with the modal damping at each mode can be obtained experimentally for any structures and machines. This process is called the experimental modal analysis. This process is useful for conducting *in-situ* modal tests on machines and structures to understand their dynamics.

In the experimental modal analysis, an external dynamic force (excitation) to the structure is applied in a controlled frequency band and simultaneously the vibration responses at a number of locations are also measured. The collected vibration data are then analyzed to extract the modal parameters, namely natural frequencies, modeshapes and modal damping. The step-by-step procedures for the modal tests, data analysis and the extraction of the modal parameters are explained through a spring-mass system. This simple concept can be applied on any complex systems and found to be useful for most practical and industrial cases. This concept is further demonstrated through a few laboratory and industrial examples.

8.1.1 IMPULSIVE LOAD USING THE INSTRUMENTED HAMMER

The concept of the modal testing is based on the resonance responses of the objects to identify their modal parameters—natural frequencies, modeshapes and modal damping. The external excitation in the broad frequency band can generates resonance at several natural frequencies of the objects within the frequency band of excitation. Either vibration shaker or the instrumented hammer with force sensor can be used for this external vibration excitation in the objects. The impulsive hammer is the simplest and quickest approach for the modal tests and often useful for the *in-situ* modal tests for the structures and machines. Hence the instrumented hammer is only discussed in this chapter. The modal test using the instrumented hammer is known as the impulse-response method. In industries, this approach is popularly known as a bump test and often force measurements are done during the test; therefore, this is not a good practice.

As the name suggests, this hammer can generate an impulsive excitation to structures and then the structures follow free decay of vibration response. In general, it appears as a normal hammer which consists of a force transducer for measuring how much force is applied to structure and an impacting head. A schematic of the instrumented hammer is shown in Figure 8.1. The photograph of a small instrumented hammer is also shown in Figure 8.2. For bigger structures and machines, a big instrumented hammer can be used as shown in Figure 8.3.

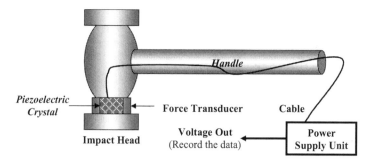

FIGURE 8.1 An instrumented hammer.

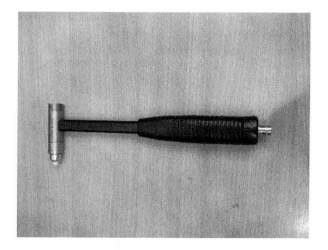

FIGURE 8.2 Photograph of a small instrumented hammer.

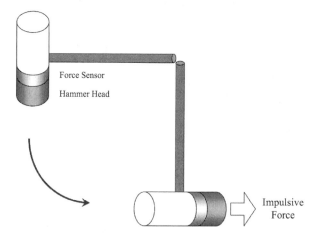

FIGURE 8.3 Pendulum arrangement used for the hammer to apply impulsive excitation to the large objects for modal testing.

Experimental Modal Analysis

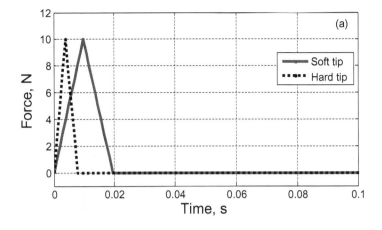

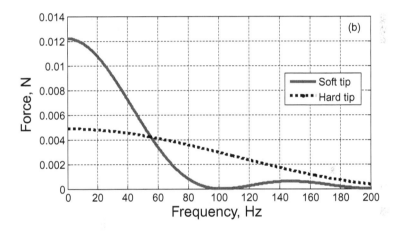

FIGURE 8.4 Affect of different impact heads of an instrumented hammer (a) time waveforms and (b) spectra of the applied forces.

Force sensor in the instrumented hammer is made of a piezoelectric crystal, which generates charge due to deformation in the crystal as a result of the impact applied. The frequency band of excitation can be controlled by the use of different impact head of the hammer. Low- to high-frequency range of excitation can be achieved by using a softer to harder impact tip, respectively, for the hammer. Figure 8.4 shows a typical comparison between two impacts by a hammer, one with soft-head tip and another with relatively harder-head tip. The applied peak forces by both head tips are exactly same as shown in Figure 8.4a, but the time length for the soft-head tip is longer as expected compared to the hard tip; therefore, the frequency excitation band for the soft-head tip is going to be lower than the hard head tip. Spectra of both impulses in Figure 8.4b clearly show this feature. The excitation force level is higher for the soft-head tip but the excitation frequency band is only upto 50–60 Hz whereas the low excitation force but high frequency band upto 150–160 Hz for the hard-tip hammer.

Therefore it is good to select the different hammer impact head tip to meet the required frequency band of excitation for the object to be tested. Instrumented hammers of different sizes, impact capacities, impact head hardness, etc. are commercially available.

8.2 MODAL ANALYSIS

To explain this method, a simple example of an SDOF system consisting of a mass, a spring with an inherent damping as shown in Figure 8.5 is considered here. Now it is assumed that the modal parameters—natural frequency (f_n), the modeshape (ϕ) and the modal damping (ζ)—are not known and need to be estimated experimentally.

Step 1: The modal testing experiment: The modal test experiment is conducted as per the measurement scheme shown in the schematic in Figure 8.5. The force and the response signals using the force sensor and the response accelerometer are collected and stored in a PC. In the presented case, the data has been collected at a sampling rate, $f_s = 256$ samples/sec. The measured force and the acceleration response are shown in Figures 8.6 and 8.7.

Step 2: Computation of the amplitude spectrum, FRF and coherence: Since the sampling frequency, f_s is 256 Hz and data collected for 8 s for both force and acceleration response, hence the number of the data points, $N = 2^{11} = 8 \times 256 = 2048$ for the FFT calculation are used. This leads to the frequency resolution, $df = \frac{f_s}{N} = \frac{256}{2048} = 0.125$ Hz, the Nyquist frequency, $f_q = \frac{f_s}{2} = 128$ Hz and the upper useful frequency, $f_u = \frac{f_s}{2.56} = 100$ Hz. The calculated amplitude spectra for the force and acceleration signals are shown in Figures 8.8 and 8.9. It is obvious from the force spectrum that the excitation level is good upto 80–90 Hz.

The FRF amplitude and coherence plots for the input force and the output response are shown in Figures 8.10 and 8.11. Coherence is nearly 1 in

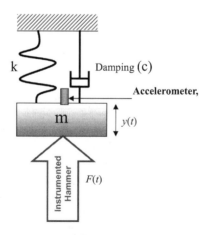

FIGURE 8.5 Modal testing on an SDOF system.

Experimental Modal Analysis

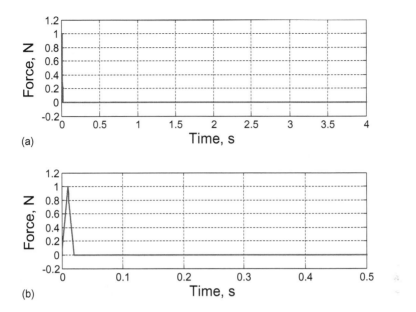

FIGURE 8.6 (a) Measured force and (b) its zoomed view in 0–0.5 s time.

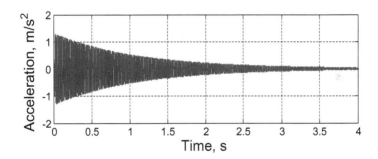

FIGURE 8.7 Measured vibration acceleration response due to applied impulsive force in Figure 8.6.

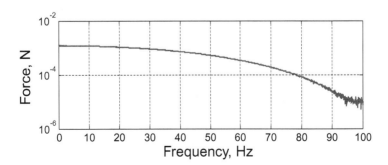

FIGURE 8.8 Averaged spectrum of the measured force.

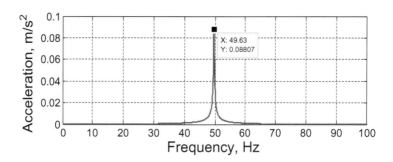

FIGURE 8.9 Spectrum of the measured acceleration response.

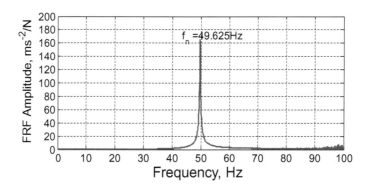

FIGURE 8.10 FRF plot of the measured acceleration response to the applied for the SDOF example.

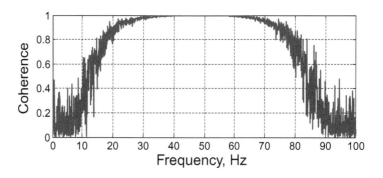

FIGURE 8.11 Coherence between the measured acceleration response and the applied for the SDOF example.

the excitation frequency range from 30 to 70 Hz confirming the fact that the force and response are linearly related and the quality of the measurements is very good between 30–70 Hz. Hence the peak at 49.625 Hz in the acceleration spectrum and the FRF plot is related to the input force excitation. This frequency may be the natural frequency of the system.

Experimental Modal Analysis

Step 3: Identification of natural frequency from the FRF plot: To identify, whether the peak seen at 49.625 Hz is a natural frequency or not, the theory of the SDOF system discussed in Chapter 2 is considered here. The equation of displacement, Equation (2.33), in the frequency domain can be rearranged as

$$\frac{y(f)}{F(f)} = \frac{1}{\left((k - m\omega^2) + jc\omega\right)} = \frac{(1/k)}{\left(1 - \left(\frac{f}{f_n}\right)^2\right) + j\left(2\zeta\frac{f}{f_n}\right)} \tag{8.1}$$

Equation (8.1) is nothing but the FRF of the system where the frequency, f is the exciting frequency and hence $F(f)$ and $y(f)$ are the applied force and the displacement response of the system at the frequency, f. f_n is the natural frequency. The real part, $(k - m\omega^2)$ in Equation (8.1) is nothing but the resultant of the inertia and stiffness force in the system and the imaginary part, $c\omega$ is related to the damping force. Since the measured response is the acceleration, hence the FRF in Equation (8.2) in terms of the acceleration can be written as

$$FRF(f) = \frac{a(f)}{F(f)} = \frac{-\omega^2 / k}{\left(1 - \left(\frac{f}{f_n}\right)^2\right) + j\left(2\zeta\frac{f}{f_n}\right)} \tag{8.2}$$

The FRF in Equation (8.2) is also known as inertance of the system. Equation (8.2) can also be written as

$$FRF(f) = \frac{-\omega^2 / k}{\left(1 - \left(\frac{f}{f_n}\right)^2\right) + j\left(2\zeta\frac{f}{f_n}\right)} \times \frac{\left(1 - \left(\frac{f}{f_n}\right)^2\right) - j\left(2\zeta\frac{f}{f_n}\right)}{\left(1 - \left(\frac{f}{f_n}\right)^2\right) - j\left(2\zeta\frac{f}{f_n}\right)}$$

$$FRF(f) = \frac{(-\omega^2 / k)\left[\left(1 - \left(\frac{f}{f_n}\right)^2\right) - j\left(2\zeta\frac{f}{f_n}\right)\right]}{\left(1 - \left(\frac{f}{f_n}\right)^2\right)^2 + \left(2\zeta\frac{f}{f_n}\right)^2} \tag{8.3}$$

Equation (8.3) can further been written as

$$FRF(f) = A(f) + jB(f) \tag{8.4}$$

where

$$\text{Real Part, } A(f) = \frac{(-\omega^2/k)\left(1-\left(\dfrac{f}{f_n}\right)^2\right)}{\left(1-\left(\dfrac{f}{f_n}\right)^2\right)^2 + \left(2\zeta\dfrac{f}{f_n}\right)^2} \tag{8.5}$$

$$\text{Imaginary part, } B(f) = \frac{(\omega^2/k)\left(2\zeta\dfrac{f}{f_n}\right)}{\left(1-\left(\dfrac{f}{f_n}\right)^2\right)^2 + \left(2\zeta\dfrac{f}{f_n}\right)^2} \tag{8.6}$$

and the phase angle between the response with respect to the applied force at frequency, f

$$\phi(f) = \tan^{-1}\left(\frac{B(f)}{A(f)}\right) = \tan^{-1}\left(\frac{\left(2\zeta\dfrac{f}{f_n}\right)}{\left(1-\left(\dfrac{f}{f_n}\right)^2\right)}\right) \tag{8.7}$$

Let's assume that the peak at the frequency 49.625 Hz in the FRF inertance plot in Figure 8.10 is a natural frequency, f_n. It is also assumed that the frequency of the applied force frequency, f is now equals to f_n then the real part, imaginary part and the phase of the FRF from Equations (8.5) through (8.7) are

$$\text{Real part, } A(f_n) = 0$$

$$\text{Imaginary part, } B(f_n) = \frac{(\omega_n^2/k)}{2\zeta}$$

and the phase at frequency, f_n,

$$\phi(f_n) = \tan^{-1}\left(\frac{B(f_n)}{A(f_n)}\right) = \tan^{-1}\left(\frac{(2\zeta)}{0}\right) =$$

$+$ or $-$ 90 degree (depending on force and response)

Hence it is clear from the above simple calculations that the FRF real part will be *zero*, imaginary part may show either positive or negative peak and phase becomes equal to 90° at the resonance (i.e., when the exciting frequency equals the natural frequency). The peak at 49.625 Hz

Experimental Modal Analysis

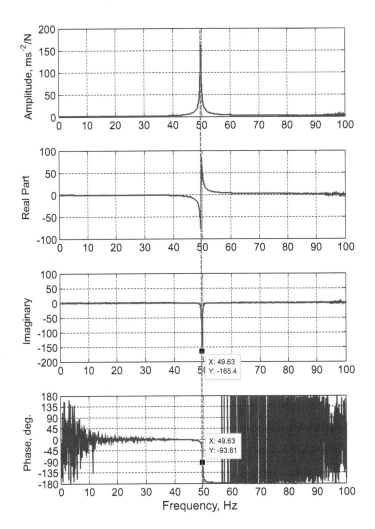

FIGURE 8.12 FRF plot showing amplitude, real, imaginary and phase.

meets all these criteria as clearly seen in Figure 8.12. Hence the frequency 49.625 Hz is the natural frequency of the SDOF system. The physical significance of the terminologies in the context of modal analysis can be summarized as follows:

1. *Real part:* It is basically the resultant of the stiffness force and the inertia force. A zero value indicates that the inertial force in the system is completely balanced by stiffness force at the natural frequency.
2. *Imaginary part:* The peak at the natural frequency shows that the applied force is completely balanced by the damping force. This means that the system completely loses its mechanical strength

168   Industrial Approaches in Vibration-Based Condition Monitoring

(Real Part = 0) at the natural frequency and the response can go to infinity in absence of any damping.

3. *Phase:* ±90° phase difference of the response with respect to the applied force at the natural frequency.

Step 4: Modeshape extraction: The amplitude at 49.625 Hz in the FRF plot (Figure 8.12) is made up of a complex number (0–165.4j). The real part is close to zero as expected at the natural frequency so it can be ignored and the imaginary part of −65.4 can be used for the modeshape for the mass movement. This simply means that the mass movement is ±165.4 (±1 on a normalized scale) in one complete cycle with the time period, $T = 1/49.625$ s. The value of 165.4 and its unit (ms^{-2}/N) is not important for the modeshape. It is because the modeshape at a natural frequency just defines the shape of deformation. Section 2.4 of Chapter 2 explains this concept of the modeshape more clearly.

Step 5: Half-power point (HPP) method for estimation modal damping: There are a number of methods to estimate the modal damping from the modal test data. However, the half-power point (HPP) method is a simple and fairly good alternative for the estimation of modal damping for most practical purpose. This is discussed here.

The amplitude of the FRF of the displacement to the force in Equation (8.1) in terms of the non-dimensional form is written as

$$Y_{ND}(f) = \frac{ky(f)}{F(f)} = \frac{1}{\sqrt{\left[1 - \left(\frac{f}{f_n}\right)^2\right]^2 + \left(2\zeta\frac{f}{f_n}\right)^2}} \tag{8.8}$$

where $Y_{ND}(f)$ is the non-dimensional amplitude of the FRF. At the natural frequency, $Y_{ND}(f_n) = \frac{1}{2\zeta}$ and in terms of power, $Y_{ND}^2(f_n) = \left(\frac{1}{2\zeta}\right)^2$ and so the half-power point will be $\frac{1}{2}\left(\frac{1}{2\zeta}\right)^2$ and the non-dimensional amplitude becomes $\frac{1}{\sqrt{2}}\left(\frac{1}{2\zeta}\right) = 0.707\left(\frac{1}{2\zeta}\right)$. Substituting this half-power point value into Equation (8.8) gives

$$\frac{1}{2}\left(\frac{1}{2\zeta}\right)^2 = \frac{1}{\left[1 - \left(\frac{f}{f_n}\right)^2\right]^2 + \left(2\zeta\frac{f}{f_n}\right)^2} \tag{8.9}$$

On simplification the solution of Equation (8.9) gives two roots for the frequency, f and they are $f_1 = (1-2\zeta)f_n$ and $f_2 = (1+2\zeta)f_n$. They are depicted in the non-dimensional power and amplitude plots of FRF in Figure 8.13. The damping ratio can then be estimated as

$$\zeta = \frac{f_2 - f_1}{2f_n} \tag{8.10}$$

Experimental Modal Analysis

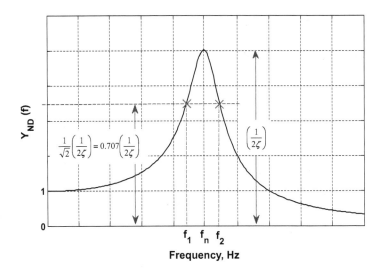

FIGURE 8.13 Non-dimensional FRF plot showing peak and half power amplitudes.

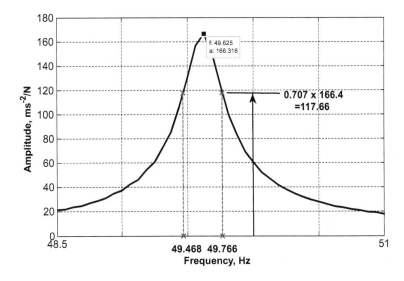

FIGURE 8.14 Zoomed view of the FRF plot in Figure 8.10.

Thus, the HPP method is used to estimate the system damping using the measured FRF amplitude plot as shown in Figure 8.10. The frequencies related to the half-power amplitude at the natural frequency, 49.625 Hz are found to be $f_1 = 49.468$ Hz and $f_2 = 49.766$ Hz. These frequencies are also marked in the zoomed view of the FRF plot in Figure 8.14. The damping for the system using Equation (8.10) is estimated as 0.003 (0.3%). Hence, the system damping is 0.3% at the natural frequency, $f_n = 49.63$ Hz.

8.3 EXPERIMENTAL EXAMPLES

8.3.1 Example 8.1—A Clamped-Clamped Beam

This is an experimental example of the modal testing on a clamped-clamped beam. Figure 8.15 shows the photograph of the experimental set-up of the beam together with clamping details. The beam span between the clamps is 2.54 m with a cross-sectional area of 30 mm (width) × 20 mm (depth). It is made up of aluminum. The physical dimensions of the beam viz. Length-width-depth are assumed to be along x-y-z of the reference axes, respectively.

Step 1: The modal testing experiment: Figure 8.16 shows the photograph of the beam with 13 accelerometers (sensitivity 100 mV/g) mounted along the length of the beam to measure the response in the z-direction (depth) (i.e., lateral direction). The instrumented hammer with the force sensor has a sensitivity of 1.1 mV/N as shown in Figure 8.16 and is used to excite the beam for the modal test. The schematic of the accelerometers locations on the beam together with associated instruments, DAQ device and the impulse hammer is shown in Figure 8.17. A number of impulses are given to excite the beam. The impulse together with the vibration responses from all 13 accelerometers are simultaneously recorded into the PC through a 16-input channels 16-bit DAQ device using the sampling frequency of 5000 Hz. A typical measured impulsive excitation force and the measured response at Location 3 are shown in Figure 8.18.

Step 2: Computation of the amplitude spectra, FRFs and coherences: Averaged acceleration spectra are calculated for both the force and the 13 responses data initially and then the FRFs and the coherences are calculated for each accelerometer data with respect to the applied force.

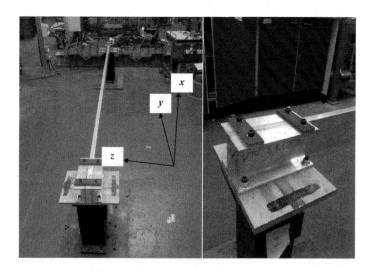

FIGURE 8.15 Example 8.1—a clamped-clamped beam.

Experimental Modal Analysis

FIGURE 8.16 Beam with accelerometers and instrumented hammer for modal testing.

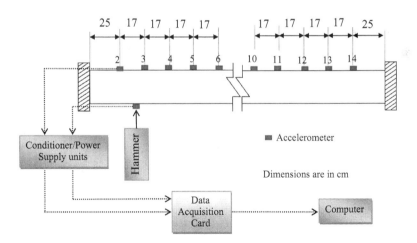

FIGURE 8.17 Schematic view of the experiment for the clamped-clamped beam.

These calculations are carried out as per Equations (5.10, 5.14, 5.15, 5.18 and 5.20) of Chapter 5. Since the applied force and the decay responses are short duration phenomena, special care is needed for the signal processing to compute the averaged spectra, FRFs and coherences. A trigger level of 1.0 N has been used as the starting point of acquisition of data

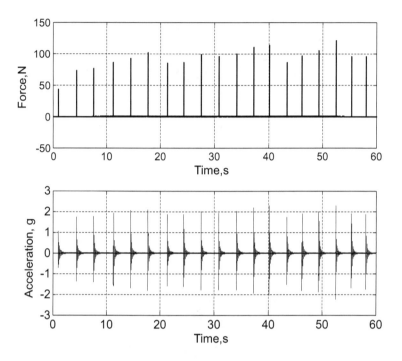

FIGURE 8.18 Typical measured data (a) applied force by instrumented hammer (b) beam acceleration response measured by an accelerometer at location 3.

for the force and the corresponding responses as shown in Figure 8.19 to compute the FT using Equation (5.10). The time length for each FFT calculation corresponds to the segment size, $N = 2^{13} = 8192$. This process is repeated for all the applied forces and the corresponding acceleration responses from all 13 accelerometers. The averaged spectra, FRF data and coherences are then calculated using Equations (5.15, 5.18 and 5.20), respectively. Typical spectra for the applied force and the acceleration response at Location 3 and their FRF and coherence plots are shown in Figures 8.20 through 8.22, respectively. The spectrum of the excitation force shows nearly constant excitation upto the frequency range of 200 Hz and there are three peaks seen in the response spectrum and in the FRF plot. Coherence at these three peaks is also observed to more than 0.8 indicates a good relation between the applied force and the measured response.

Step 3: Natural frequencies identification from the FRF plots: Now the three distinct peaks in the response spectrum and the FRF plot in Figures 8.20 and 8.21 are assumed to be the natural frequencies for the experimental beam to begin the identification process. The frequency values for these three peaks are $f_{p1} = 17.70$ Hz, $f_{p2} = 54.9316$ Hz and $f_{p3} = 103.76$ Hz. Now each peak is analyzed based on the theoretical concept of the SDOF system and hence the following examinations need to be done (as discussed in Section 8.3

Experimental Modal Analysis

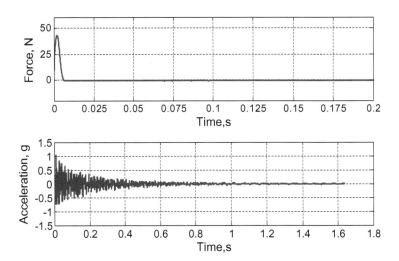

FIGURE 8.19 Trigger force and the corresponding response at location 3 used for the FFT estimation for each applied force for averaging purpose.

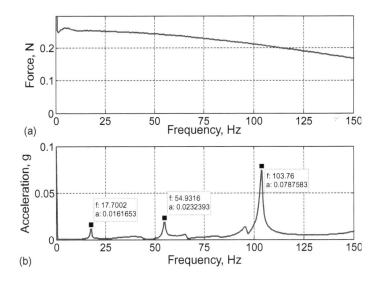

FIGURE 8.20 Averaged amplitude spectra of the applied force (a) and the beam response (b) shown in Figure 8.18.

(Step 3) for the each peak at all the 13 measurement locations using the FRF plots to confirm whether it is a natural frequency or not.

1. Real part must pass through zero.
2. Imaginary part must be either a +ve or −ve peak.
3. Phase angle of the measured response with respect to the applied force at the frequency peak must pass through ±90 degree.

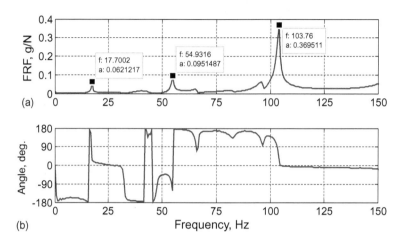

FIGURE 8.21 Measured FRF (inertance) at the location 3 (a) FRF amplitude (b) FRF phase.

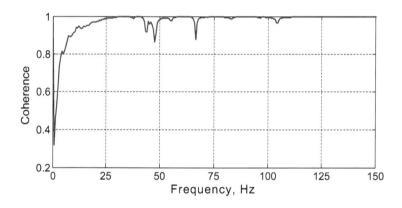

FIGURE 8.22 Coherence between the applied and the response shown in Figure 8.18.

Majority of the measured locations must meet the above criteria except when the measurement location is close to a node (zero deflection) location for any natural frequency/frequencies. It is to be noted that this approach is a very simple yet elegant procedure for the extraction of modal parameters from the experimental data, and at the same time quick and easy in assessment approach and generally works well for most of the cases.

The FRF at each measured location is examined for the above criteria for each frequency peaks and it is observed that all these three frequencies satisfy the criteria to be the natural frequencies. Few typical FRF plots are also shown in Figures 8.23 through 8.25 for better clarity. Hence the frequencies are now identified as the first three natural frequencies in the lateral direction of the experimental beam.

Experimental Modal Analysis

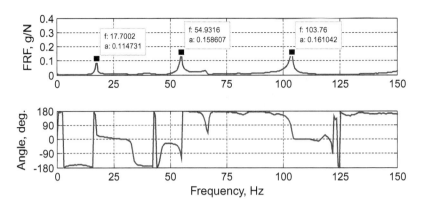

FIGURE 8.23 Measured FRF (inertance) at the location 6.

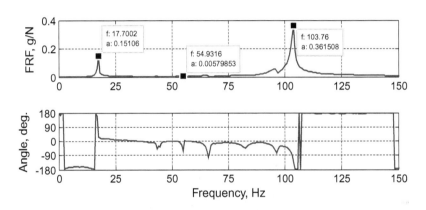

FIGURE 8.24 Measured FRF (inertance) at the location 8 (beam centre).

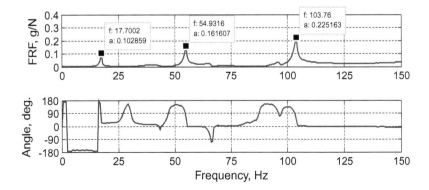

FIGURE 8.25 Measured FRF (inertance) at the location 11.

TABLE 8.1

Measured Imaginary Values at The Natural Frequencies and The Normalized Modeshapes

Measurement Location		Imaginary Values from Measured FRFs			Normalized Modeshapes		
Position	Distance, cm	φ_{I1}	φ_{I2}	φ_{I3}	φ_1	φ_2	φ_3
1	0	0	0	0	0	0	0
2	25	+0.0156	−0.0211	+0.1534	+0.2367	−0.1509	0.5713
3	42	+0.0261	−0.0784	+0.2396	+0.3961	−0.5608	0.8924
4	59	+0.0385	−0.1252	+0.2313	+0.5842	−0.8956	0.8615
5	76	+0.0492	−0.1353	+0.1139	+0.7466	−0.9678	0.4242
6	93	+0.0573	−0.1195	−0.0550	+0.8695	−0.8548	−0.2048
7	110	+0.0633	−0.0718	−0.1978	+0.9605	−0.5136	−0.7367
8	127	+0.0659	−0.0031	−0.2492	+1.0000	−0.0222	−0.9281
9	144	+0.0637	+0.0664	−0.1783	+0.9666	0.4750	−0.6641
10	161	+0.0580	+0.1208	−0.0183	+0.8801	0.8641	−0.0682
11	178	+0.0461	+0.1398	+0.1534	+0.6995	1.0000	0.5713
12	195	+0.0368	+0.1387	+0.2685	+0.5584	0.9921	1.0000
13	212	+0.0247	+0.1015	+0.2600	+0.3748	0.7260	0.9683
14	229	+0.0133	+0.0491	+0.1459	+0.2018	0.3512	0.5434
15	254	0	0	0	0	0	0

Step 4: Extraction of modeshapes: As discussed earlier, the imaginary values of the FRF plot at each measured location for each natural frequency can be used for generating the modeshape. The modeshapes of the three identified natural frequencies directly measured from the imaginary part of the FRF plots are listed in Table 8.1. The normalized values of the modeshapes are also listed in Table 8.1.

It is to be noted that there are no measurements at the clamped locations (1 and 15) on either side of the beam. So it is assumed that there are no deflections at the clamped locations. The modeshapes data (Table 8.1) in terms of the imaginary values from the measured FRFs are denoted as vectors $\varphi_I = \begin{bmatrix} \varphi_{I1} & \varphi_{I2} & \varphi_{I3} \end{bmatrix}$ as the modeshapes for the first (Mode 1), second (Mode 2) and third (Mode 3), respectively. Since the modeshape is the shape of deformation at each mode for the structure, so the measured values are not important, hence the modeshapes are normalized between ±1 by dividing modeshape data at a natural frequency by its absolute maximum value. For the present case the normalized modeshape data for the three modes are estimated as $\varphi = \begin{bmatrix} \frac{\varphi_{I1}}{0.0659} & \frac{\varphi_{I2}}{0.1398} & \frac{\varphi_{I3}}{0.2685} \end{bmatrix} = \begin{bmatrix} \varphi_1 & \varphi_2 & \varphi_3 \end{bmatrix}$ which are also listed in Table 8.1. The measured modeshapes for Mode 1, Mode 2 and Mode 3 are also shown graphically in Figure 8.26. The modal assurance criteria (MAC) are also calculated between the experimental modes using Equation (8.11).

Experimental Modal Analysis

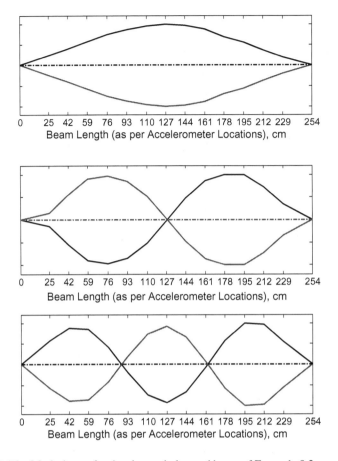

FIGURE 8.26 Modeshapes for the clamped-clamped beam of Example 8.2.

$$MAC_{ij} = \frac{\left|\sum_{k=1}^{n}\phi_{k,i}\phi_{k,j}\right|^2}{\left(\sum_{k=1}^{n}\phi_{k,i}\phi_{k,i}\right)\left(\sum_{k=1}^{n}\phi_{k,j}\phi_{k,j}\right)} \tag{8.11}$$

where "n" is the number of measurement locations, $(\varnothing_{k,i}, \varnothing_{k,j})$ are the modeshape at location k for ith and jth modes, respectively. The value of MAC indicates the relation between the two modes. The MAC close to 1 means modes are 100% related and 0 means no correlation (i.e., both modes are orthogonal to each other). The estimated MAC values are shown in the 3-D bar graph in Figure 8.27 and also tabulated in Table 8.2. It is obvious from Figure 8.27 and Table 8.2 that the cross-coupling Mode 1 with Modes 2 and 3, and between the Modes 2 and 3 are nearly negligible, hence the modes are orthogonal to each other.

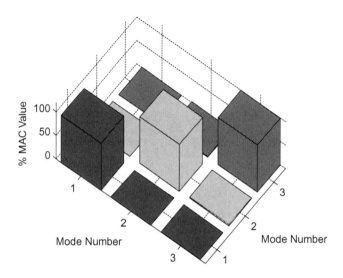

FIGURE 8.27 Bar graph for % MAC values.

TABLE 8.2
MAC Values (%) Between the Modes of Example 8.2

	Mode 1	Mode 2	Mode 3
Mode 1	100	0.0067	0.1187
Mode 2	0.0067	100	1.8183
Mode 3	0.1187	1.8183	100

Step 5: Estimation modal damping: Here again, the HPP method is used to compute the modal damping ratio at all the 13 measurement location for Mode 1 to Mode 3 using the measured FRF inertance plots. The estimated damping values are listed in Table 8.3. The estimated damping ratios at each mode at all the 13 measured locations may not be exactly same due to different noise content and other factors at each measurement location. A few measurement locations for Mode 2 and Mode 3 are either at the node locations or close to the nodal locations may show significantly high damping as listed in Table 8.3. These values should be ignored. Therefore it is always advisable to choose the lowest damping ratio at each mode as a conservative approach. Hence the measured damping ratios at Mode 1 to Mode 3 are 1.494%, 0.716% and 0.592%, respectively. These values are also shown in Table 8.3.

8.3.2 Example 8.2—Experimental Rotating Rig-1

The experimental rig-1 (Shamsah et al., 2019) shown in Figure 8.28 is consists of a mild steel solid shaft (1000 mm length and 20 mm diameter) supported on flexible

TABLE 8.3
Damping Ratio Estimated by HPP Method

Measurement Location	Modal Damping Ratios (%)		
	Mode 1	Mode 2	Mode 3
2	1.4940 (L)	0.7594	0.6161
3	1.4991	0.7223	0.6198
4	1.4986	0.7193	0.6258
5	1.4988	0.7190	0.6504
6	1.4986	0.7203	0.5916 (L)
7	1.5021	0.7294	0.6114
8	1.5019	21.7569 (NL)	0.6233
9	1.5062	0.7194	0.6468
10	1.5085	0.7162 (L)	3.7727 (NL)
11	1.5178	0.7177	0.5918
12	1.5279	0.7187	0.6102
13	1.5444	0.7234	0.6243
14	1.5773	0.7258	0.6344
Final (lowest) Damping	1.4940	0.7162	0.5916

L for lowest value, NL refers a proximity to node location

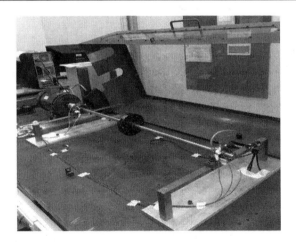

FIGURE 8.28 Experimental rotating rig-1.

foundations through two ball bearings The rotor shaft is connected to an electric motor through a flexible coupling The rotor is also carrying a balance disc.

The impulse-response method is used for the modal tests at 0 speed of the rig (Shamsah et al., 2019). The measured FRF plots at a location in the vertical and horizontal directions are shown in Figure 8.29. The peaks at frequencies 17.09 Hz, 29.91 Hz, 31.13 Hz and 58.59 Hz in the FRF plots generally meet the criteria of the FRF real, imaginary and phase

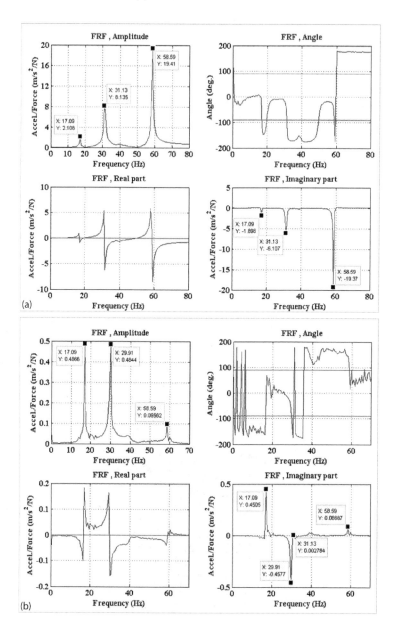

FIGURE 8.29 Modal tests: Measured FRF plots at a distance of 42 cm from bearing near motor for the experimental rig-1 in (a) vertical and (b) horizontal directions.

at all measurement locations. Hence the frequencies 17.09 Hz, 29.91 Hz, 31.13 Hz and 58.59 Hz are the natural frequencies of the rig. The corresponding modeshapes in both vertical and horizontal directions are shown in Figure 8.30. The modeshape at 29.91 Hz in Figure 8.30b is not showing any deformation in the vertical direction so this mode is predominantly mode in the horizontal direction only.

Experimental Modal Analysis

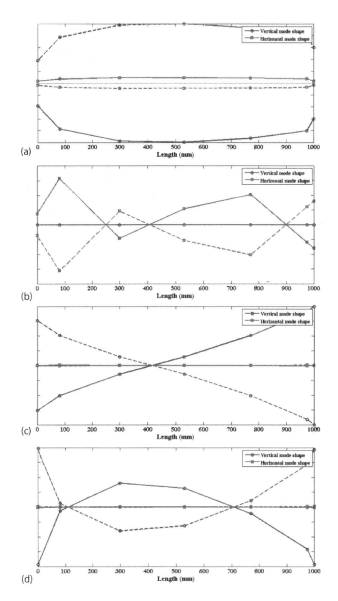

FIGURE 8.30 Measured modeshapes of the experimental rig-1 (a) Mode 1: 17.09 Hz (b) Mode 2: 29.91 Hz (c) Mode 3: 31.13 Hz and (d) Mode 4: 58.59 Hz.

8.3.3 EXAMPLE 8.3—EXPERIMENTAL ROTATING RIG-2

Figure 8.31 shows a photograph of the experimental rig-2. The rig consists of two 20 mm nominal diameter rigidly coupled (Coupling 2) of length shafts of 1010 mm (Rotor 1) and 500 mm (Rotor 2). The complete rotor is supported on a steel lathe bed through four ball bearings. The rig rotor is connected to an electric motor through a flexible coupling (Coupling-1).

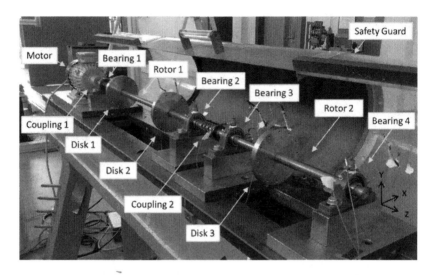

FIGURE 8.31 Experimental rotating rig-2.

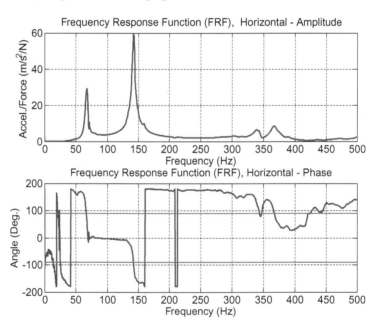

FIGURE 8.32 Modal tests: Measured FRF on the shaft near to bearing B2 for the experimental rig-2 in the horizontal direction.

Modal tests are carried out in the horizontal direction using the impulse response method to identify the natural frequencies and their modeshapes (Nembhard and Sinha, 2015). A typical measured FRF plots (amplitude and phase) of the measured acceleration response to the applied impulsive force are shown in Figure 8.32. The first two natural frequencies in the FRF satisfy the criteria for the natural

Experimental Modal Analysis

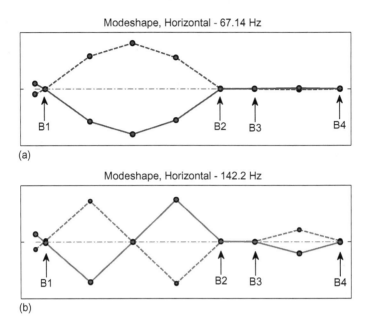

FIGURE 8.33 Measured modeshapes (dots represent measurement locations) of the experimental rig-2 in horizontal direction only (a) Mode 1: 67.14 Hz and (b) Mode 2: 142.2 Hz.

frequencies. Hence these two frequencies at 67.14 Hz and 142.2 Hz are identified the natural frequencies—Mode1 and Mode 2. The modeshapes corresponding to these natural frequencies are also provided in Figure 8.33.

8.4 INDUSTRIAL EXAMPLES

8.4.1 Example 8.4—Horizontal Centrifugal Pump

This is a typical case study of an industrial centrifugal pump commissioned in 1985 (Sinha and Rao, 2006). This pump had never experienced any frequent failure until 2004, and then suddenly began to experience frequent failures of its anti-friction bearing. The vibration-based condition monitoring has always detected the appearance of bearing faults in advance and this information has always been used to initiate a well-planned shutdown for the replacement of the faulty bearing. However the condition monitoring could not identify the root cause for this frequent failure of the bearings. The modal test is found to be a tool to identify the root cause and then solve the problem.

Figure 8.34 shows the schematic of the pump assembly. It is a horizontally mounted centrifugal pump with an axial inlet, and a radial outlet. The pump and the motor shafts are rigidly coupled to a shaft carrying a flywheel (FW). The FW is supported by a grease lubricated radial bearing on the pump side and oil lubricated taper roller thrust bearing on the other side. The pump is driven by a 540 kW electric

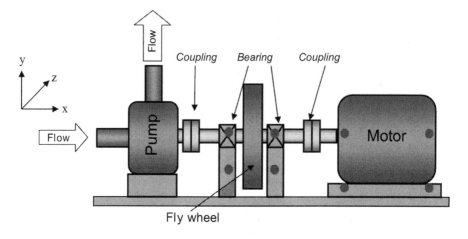

FIGURE 8.34 Schematic of a pump assembly and the measurement (dot) locations.

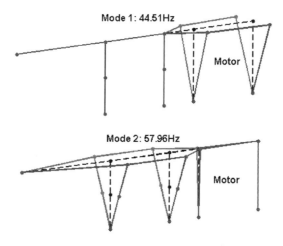

FIGURE 8.35 Modeshapes at natural frequencies 44.51 Hz and 57.96 Hz of the horizontal pump assembly.

motor operating at 1492 RPM. The pump is mounted directly on the base plates embedded into a rigid concrete floor.

The modal test conducted on the complete pump assembly by Sinha and Rao (2006). The modeshapes of the identified natural frequencies are shown in Figure 8.35. The modeshape at the natural frequency 57.96 Hz had significant deflection only at both bearing pedestals (Figures 8.35 and 8.36). This can be clearly visualized from Figure 8.36. This frequency in the FRF at the bearing pedestals appeared as a broad banded peak and has almost 20 dB amplification at 2 × component of the pump vibration. It is typically seen in the frequency response function (FRF) plot in z-direction (lateral to the rotor axis) at the bearing pedestal

Experimental Modal Analysis

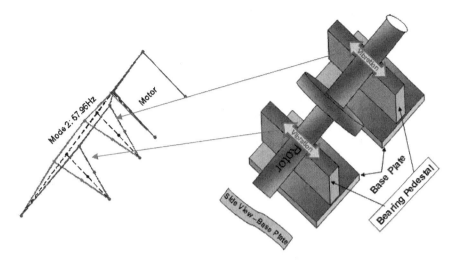

FIGURE 8.36 Modeshape at 57.96 Hz together with the pump assembly highlights the deflection of bearing pedestals only.

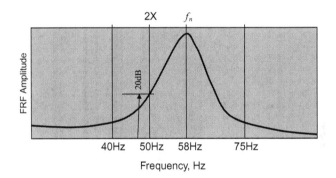

FIGURE 8.37 A typical FRF plot at bearing pedestal (near to motor) in lateral direction (z-direction).

(near the motor side) in Figure 8.37. The drop in the natural frequency close to the 2×, and its broadbanded nature must be due to the looseness between the base plate and the concrete, resulting in the non-linear interaction between the base plate and the concrete surface.

Hence, the experimentally identified broadbanded natural frequency at 57.96 Hz and its closeness to 2× component has been identified as the main reason for the failure. A small 2× component generated as a result of even a smallest shaft misalignment at the coupling must have triggered the resonance at 57.96 Hz, which in turn leads to an increase in the shaft misalignment as the pump continues to operate under this condition. Such induced misalignment could cause damage to the bearings prematurely. It is a typical age-related problem of the machine foundation, which can be solved by either stiffening the roots of the bearing pedestals as

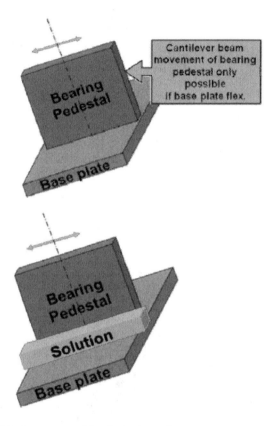

FIGURE 8.38 Solution suggested for the horizontal pump.

illustrated in Figure 8.38 or by properly grouting the base plate in concrete. Details of the vibration measurements, data analysis, diagnosis and the recommended solutions are provided in Sinha and Rao (2006).

8.4.2 Example 8.5—Vertical Centrifugal Pump

Once again the *in situ* modal tests have used to solved the high vibration problem found during the installation and commissioning of the vertical centrifugal pumps in 1991 (Sinha, 2006, Rao et al., 1997).

Figure 8.39 shows the schematic of the layout of the pumps' locations and piping. The pump assembly is shown in Figure 8.40a. A vibration survey during the commissioning showed acceptable vibration on top of the motors (i.e., farthest location from the pump base), but high vibration on the pump casing for four pumps. The direction of high vibration was specific for each pump: N-S for Pumps 2 and 4, and E-W for Pumps 1 and 5. Pump 3 had low vibration in both directions. It may be noted that the inlet and outlet piping are similar for Pumps 2 and 4, and for Pumps 1

Experimental Modal Analysis 187

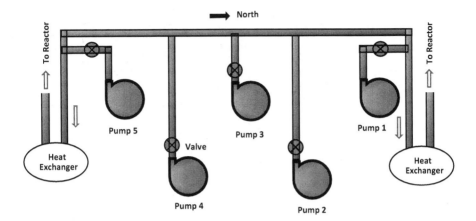

FIGURE 8.39 Schematic layout of pumps and piping.

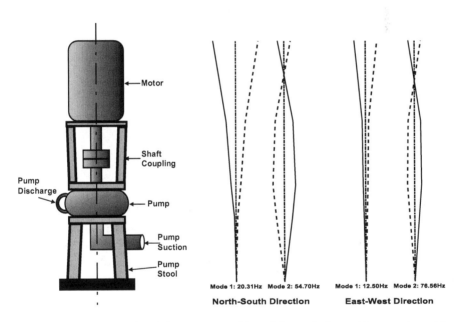

FIGURE 8.40 Schematic of the pump assembly and its modeshapes (a) pump Assembly (b) modeshapes.

and 5. Hence, these pumps cannot be considered to be safe to operate under normal conditions.

The *in situ* modal tests were conducted on all pump assemblies. The typical modeshapes at the first and second modes in orthogonal directions for Pump 2 are shown in Figure 8.40b. The second mode of the Pump 2 is found to be close to the pump

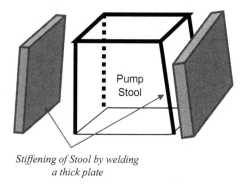

Stiffening of Stool by welding a thick plate

FIGURE 8.41 Stiffening of pump stool.

RPM and the modeshapes displayed anti-node near the pump casing. Hence, this confirms the occurrence of the resonance, which caused high casing vibration during pump operation. Similar observations were also made for remaining three pumps (Pumps 1, 4 and 5).

The natural frequency close to 3000 RPM (50 Hz) for the pump assemblies is changed by adding stiffeners to the pump stool (by welding a thick plate on each side of the stool), as shown by Figure 8.41. The support to the discharge piping of the pump was also strengthened with additional U bolts. With these modifications, the second natural frequency moved away from the pump RPM and hence the casing vibration was reduced significantly (Rao et al., 1997).

8.4.3 Example 8.6—Wind Turbine

This is an example of a typical 2.3 MW offshore wind turbine with monopile foundation (Asnaashari et al., 2018). The schematic of the wind turbine is shown in Figure 8.42 where the monopile is connected to a conical transition piece (TP). The tower is installed on top of the TP through a bolted flange connection to support the rotor-nacelle assembly.

In situ modal tests using impulse excitations were carried out by Asnaashari et al. (2018) on the wind turbine when the rotor blades were not rotating. The impulsive excitation was applied to the structure from inside the transition piece using an instrumented hammer (see Figure 8.42). Low-frequency range accelerometers are installed on the inner wall of the transition piece in the x- and y-directions as shown in Figure 8.42. The FRF plots of the measured vibration acceleration responses to the applied forces in both directions are shown in Figure 8.43. The identified natural frequencies are listed in Table 8.4.

Experimental Modal Analysis

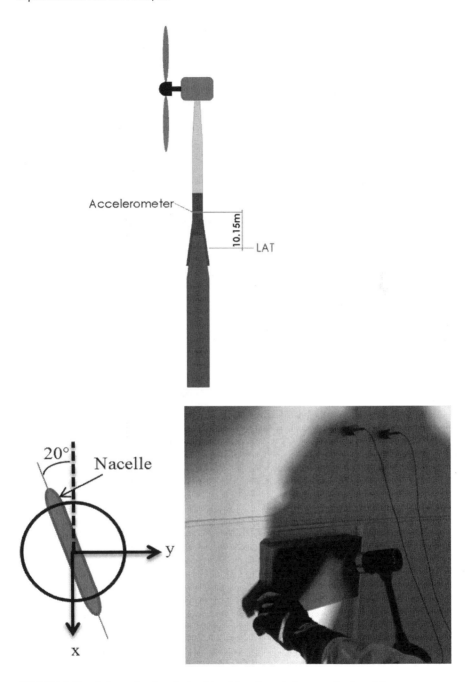

FIGURE 8.42 Schematic of a wind turbine, Nacelle and photograph of modal tests on tower.

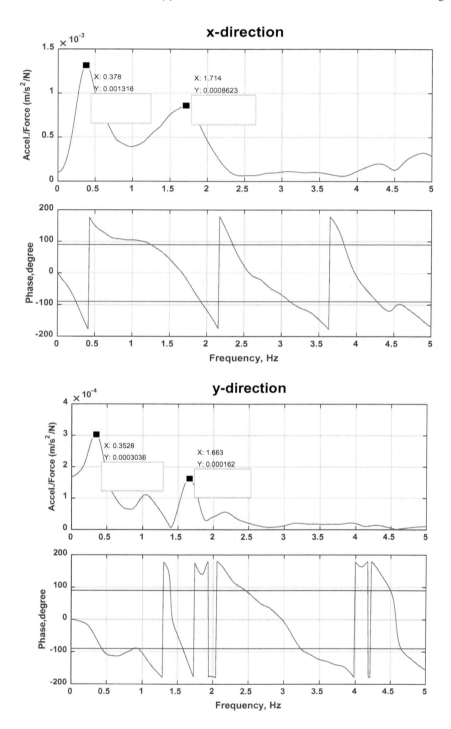

FIGURE 8.43 Measured FRF plots obtained from measured acceleration and applied impulsive forces in *x*-direction and *y*-direction.

Experimental Modal Analysis

TABLE 8.4
Experimentally Identified Natural Frequencies of a Wind Turbine

Modal Tests	x-direction	y-direction
1st bending mode, Hz	0.378	0.352
2nd bending mode, Hz	1.714	1.663

8.5 SUMMARY

A simple approach for the *in situ* modal test and analysis is presented to identify the natural frequencies, modeshapes and modal damping for any objects. The suggested method just used the theoretical of concept of an SDOF system. Each frequency peak seen in the FRF for any real system is considered to be equivalent to an SDOF system during the mode identification process. The complete procedure is explained step by step through a simple example of SDOF system and then a number of experimental and industrial examples are also presented. The importance of the modal tests in solving the machine vibration problems are also demonstrated through a couple of industrial examples.

REFERENCES

Asnaashari, E., A. Morris, I. Andrew, W. Hahn, Jyoti K. Sinha, 2018. Finite element modelling and in situ modal testing of an offshore wind turbine, *Journal of Vibration Engineering & Technologies*, 6(2), 101–106.

Nembhard, A.D., J.K. Sinha, 2015. Unified multi-speed analysis (UMA) for the condition monitoring of aero-engines, *Mechanical Systems and Signal Processing*, 64–65, 84–99.

Rao, R.A., J.K. Sinha & R.I.K. Moorthy, 1997. Vibration problems in vertical pumps—need for integrated approach in design and testing, *Shock and Vibration Digest*, 29(2), 8–15.

Shamsah, S.M.I., J.K. Sinha, P. Mandal, 2019. Estimating rotor unbalance from a single run-up and using reduced sensors, *Measurement*, 136, 11–24.

Sinha, J.K. 2006. Significance of vibration diagnosis of rotating machines during installation and commissioning: A summary of few cases, *Noise and Vibration Worldwide*, 37(5), 17–27.

Sinha, J.K., A.R. Rao, 2006. Vibration based diagnosis of a centrifugal pump, *Structural Health Monitoring: An International Journal*, 5(4), 325–332.

9 Operational Deflection Shape (ODS)

9.1 SIMPLE THEORETICAL CONCEPT

It is well known that the natural frequencies (or critical speeds) are dependent on the machine operating speeds due to gyroscopic effect for many machines. This makes it is difficult to find the critical speeds of machine experimentally due to difficult-to-conduct modal tests on any machine during its operation. The operational deflection shape (ODS) at the machine RPM and its harmonics is an alternative approach that can be used to explain the machine dynamics behavior. This analysis is often used to understand the machine vibration problems, particularly to find the reasons for frequent failures of any component (such bearing, gearbox, etc.) in machines. The ODS at a speed (or frequency) for any machine when operating is generally influenced by the modeshape of the natural frequency close to that frequency or influenced by combination of different natural frequencies close to that frequency. Following are the steps to obtain the ODS at the machine RPM or its harmonics.

A rotating machine as shown in Figure 9.1 is considered here. The machine rotating speed is 1500 PPM (25 Hz). The objective is to know the deformation shapes of the machine at 25 Hz and 50 Hz (second harmonics) when the motor is operating at 1500 RPM. It is also assumed that the natural frequencies—21.98 Hz (Mode 1) and 64.26 Hz (Mode 2)—and their modeshapes in the vertical direction are known. The modeshapes are shown in Figure 9.2. It is obvious from the modeshapes that the bearings B1–B4 are moving in phase for Mode 1 whereas B1–B3 moves in phase but B4 out of phase with a very small displacement for Mode 2.

Step 1: Here a total of four accelerometers are mounted in the vertical direction on all the bearing housings (one accelerometer per bearing) in the machine. The use of a number of accelerometers and their directions are generally depends on the requirements. For the purpose of demonstration and explain the concept of the ODS, only four accelerometers in vertical direction are used.

Step 2: Vibration measurements are carried out when machine is operating at 1500 RPM (25 Hz). The measured vibration acceleration responses at the bearings B1 to B4 are shown in Figure 9.3.

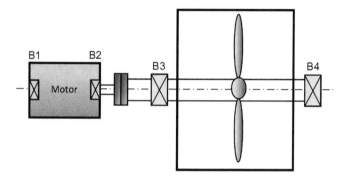

FIGURE 9.1 A typical machine mounted on flexible supports.

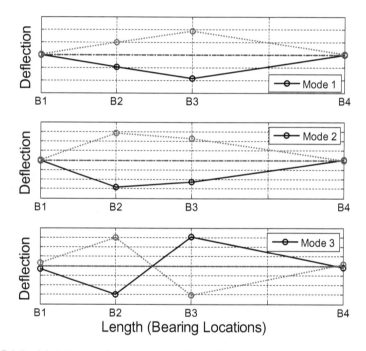

FIGURE 9.2 Modeshapes of the machine in Figure 9.1.

Step 3: Spectrum analysis is then carried out for the measured signals. The computed acceleration spectra are shown in Figure 9.4. The measured amplitudes at 25 Hz (1×) and 50 Hz (2×) from the spectrum plots for all measured locations are listed in Table 9.1.

Step 4: Since spectrum analysis only provides the amplitude of deflection at each location but no information about the phase of each location with respect to other locations. Therefore, it is impossible to draw the ODS based on the spectrum analysis only for a machine at any frequency during its operation.

Operational Deflection Shape (ODS)

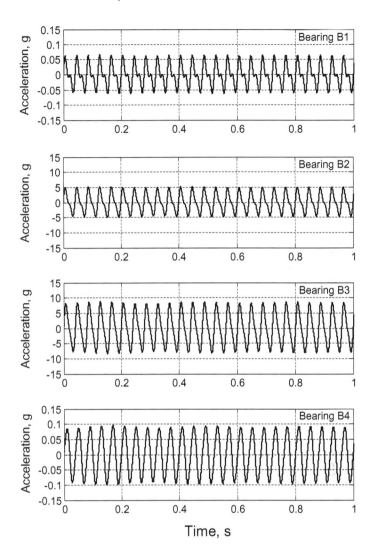

FIGURE 9.3 Measured vibration acceleration responses at bearings B1 to B4 in vertical direction.

It is essential to know the phase information with a common reference, and it is necessary to select an appropriate measurement location as a reference location to find out the relative phase of each location at a frequency with respect to the reference location. This information can be achieved by the FRF analysis. Estimate the FRF (Equation 5.18) at each location with respect to the reference location using the Equation (9.1)

$$FRF(f_k)_{mn} = \frac{S_{y_m y_n}(f_k)}{S_{y_n y_n}(f_k)} \qquad (9.1)$$

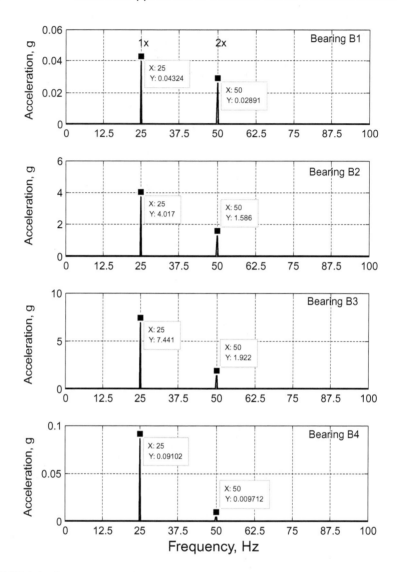

FIGURE 9.4 Spectrum plots of measured vibration shown in Figure 9.3.

where $FRF(f_k)_{mn}$ is the FRF of the signal, y_m with respect to the reference signal, y_n at the frequency, f_k.

Consider the location i.e., $n = 1$ as the reference location (Bearing B1). Then the FRF is estimated for each measurement location with respect to the reference location, B1. The estimated FRF (phase only) plots are shown in Figure 9.5.

TABLE 9.1
ODS Values at 1× and 2×

	Amplitude A, g	FRF Phase with Respect to B1 (θ), Degree	Complex Quantity, $\varnothing_m = Ae^{j\theta}$	Real Values ($\varnothing_n$ Estimated using Equation (9.2))
25 Hz (1×)				
B1	0.043249	0	0.043239	0.043239
B2	4.0171	−0.25670	4.0170−0.017997j	4.0172
B3	7.4408	−0.39726	7.4406−0.051590j	7.4409
B4	0.091020	−0.80415	0.091011−0.0012774j	0.091019
50 Hz (2×)				
B1	0.028907	0	0.028907	0.028907
B2	1.5864	−0.32140	1.5864−0.008899j	1.5865
B3	1.9222	−0.69629	1.9221−0.023359j	1.9223
B4	0.009712	−177.24	−0.00970−0.0004681j	−0.0096962

Step 5: The phase at each location with respect to the reference location B1 at 25 Hz and 50 Hz are also listed in Table 9.1. As seen from Table 9.1, the relative angle at each location is generally not exactly 0 (moving in phase with the reference location) or 180 degree (moving out of phase) at 25 Hz and 50 Hz. This is due to the presence of damping or influence of vibration from other frequencies in the system; therefore, it is essential to convert the amplitude and phase at each frequency into the real number using the following Equation (9.2) (Friswell and Mottershead, 1999).

$$\varnothing_n\left(f_k\right) = \varnothing_{m,r}\left(f_k\right) + \varnothing_{m,i}\left(f_k\right)\left[\varnothing_{m,r}^T\left(f_k\right)\varnothing_{m,r}\left(f_k\right)\right]^{-1}\varnothing_{m,r}^T\left(f_k\right)\varnothing_{m,i}\left(f_k\right) \quad (9.2)$$

where $\varnothing_m\left(f_k\right) = \varnothing_{m,r}\left(f_k\right) + j\varnothing_{m,i}\left(f_k\right)$ is the measured ODS at the frequency, f_k. $\varnothing_{m,r}\left(f_k\right)$ and $\varnothing_{m,i}\left(f_k\right)$ are the real and imaginary components of the measured ODS, $\varnothing_m\left(f_k\right)$. $\varnothing_n\left(f_k\right)$ is the normalized ODS of the complex measured ODS $\varnothing_m\left(f_k\right)$.

The estimated real numbers are also listed in Table 9.1, which are nothing but the normalized ODS, $\varnothing_n\left(f_k\right)$. Hence the ODSs at 25 Hz and 50 Hz using the normalized real numbers are plotted, which are shown in Figure 9.6. The ODS at 1× (25 Hz) is having exactly same deformation shape as the Mode 1 (21.98 Hz) modeshape; however the ODS at 2× (50 Hz) is close to Mode 2 modeshape. This means that the vibration of the machine at 50 Hz is more influenced by the second mode at 64.26 Hz.

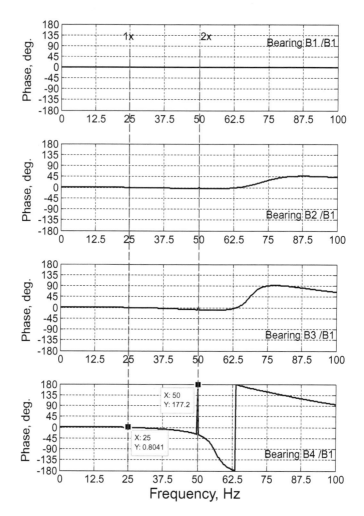

FIGURE 9.5 FRF phase plots of bearings B1 to B4 vibration responses with respect to B1 response.

9.2 INDUSTRIAL EXAMPLES

9.2.1 Example 9.1—Steam Turbo-Generator (TG) Set

Figure 9.7 is the photographs of a typical steam turbine turbo-generator (TG) set. This TG unit consists of a high pressure (HP) turbine, an intermediate pressure (IP) turbine and three low pressure (LP) turbines together with a generator and an exciter (Sinha et al., 2012). A simple schematic of TG set is shown in Figure 9.8, where B1–B14 are the fluid bearings.

The vibration measurements are done at the bearings B1 and B12 in the vertical directions during the machine rundown. The STFT analyses are carried out for

Operational Deflection Shape (ODS)

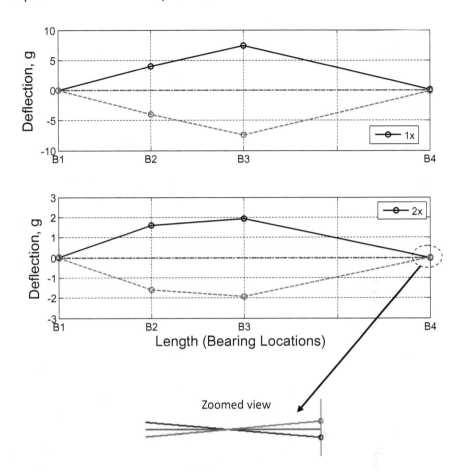

FIGURE 9.6 ODSs at 1× and 2× for the machine shown in Figure 9.1.

the measured vibration data to findout the critical speeds for the TG set. The frequencies 27.34 Hz, 33.0 Hz, 40.50 Hz and 46.88 Hz are identified as the natural frequencies (critical speeds) based on the resonance amplification in vibration when passing through critical speeds. A typical STFT plot (in contour form) during the run-down speed from 30 Hz (1800 RPM) to nearing 3000 RPM (50 Hz) is shown in Figure 9.9. The vibration amplitude amplification zones near the critical speeds at 33.0 Hz, 44.50 Hz and 46.88 Hz are clearly observed in the STFT plot.

The extracted ODSs of the bearing pedestals B1 to B12 in the TG set at the critical speeds 33 Hz and 46.88 Hz from the measured vibration during the machine rundown are typically shown in Figure 9.10. The ODSs at the critical speeds are nothing but the modeshapes at those critical speeds.

The ODS at the TG operating speed of 3000 RPM (50 Hz) is also shown in Figure 9.11. The ODS at 50 Hz is clearly close to the ODS at the critical speed of 46.88 Hz. Hence the machine dynamics of the TG set at normal operating condition is very much influenced by the critical speed at 46.88 Hz.

FIGURE 9.7 Photographs of a TG set showing (a) HP & IP turbines and (b) 2 LP turbines with vibration measurement setup.

Operational Deflection Shape (ODS)

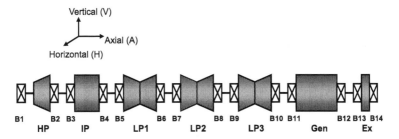

FIGURE 9.8 Schematic of the TG set.

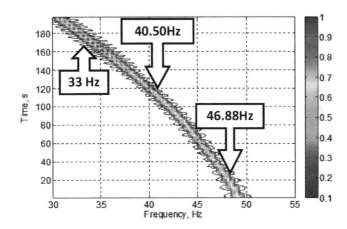

FIGURE 9.9 The measured STFT contour vibration acceleration plot at a bearing in the vertical direction during turbine run-down of the TG set.

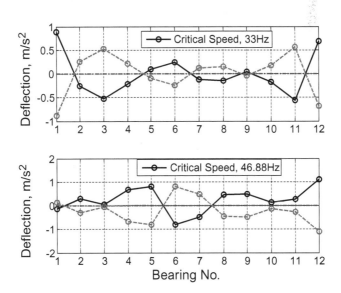

FIGURE 9.10 ODSs at the critical speeds 33 Hz and 46.88 Hz of the steam TG set.

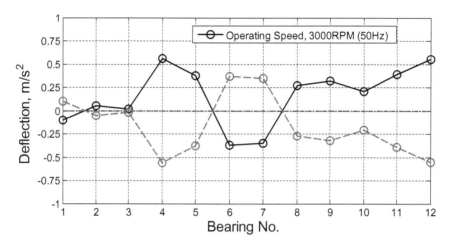

FIGURE 9.11 ODS at the operating speed of 3000 RPM (50 Hz) of the steam TG set.

9.2.2 Example 9.2—Gearbox Failure

An industrial example of the motor-gearbox-fan unit in Chapter 7 of the gear failure is again considered here to discuss the findings of the root cause of the failure through the ODS analysis (Elbhbah, Sinha and Hahn, 2018). The schematic of the unit is shown in Figure 9.12. The gearbox details are given in Figure 7.31 and the related speeds (frequencies) in Table 7.6 of Chapter 7.

The vibration measurements are done on the steel frame using a number of accelerometers in both vertical and horizontal directions (see Figure 9.12) during machine normal operation. The ODSs of the steel frame at the motor speed (24.75 Hz); the

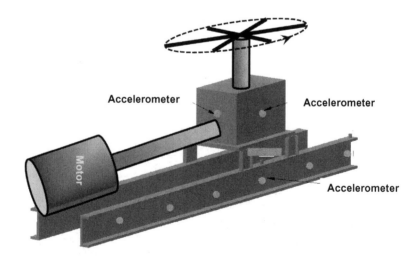

FIGURE 9.12 Schematic of the motor-gearbox-fan unit with supporting beam frame together with number of accelerometers (dots) mounted on both legs of the frame and the gearbox.

Operational Deflection Shape (ODS)

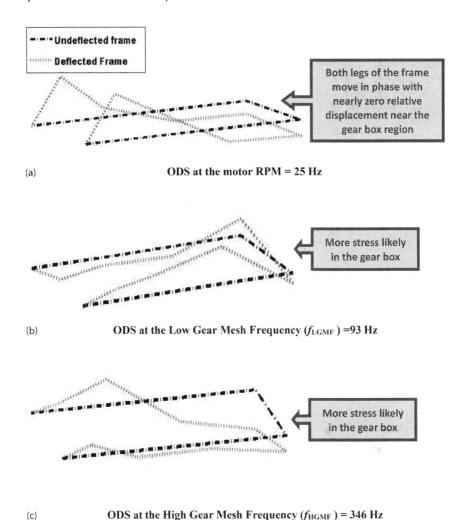

FIGURE 9.13 (a–c) ODSs of the motor-gearbox-fan supporting beam frame.

low gear mesh frequency $(f_{LGMF} = 92.4 \text{ Hz})$ and at the high gear mesh frequency $(f_{HGMF} = 346.5 \text{ Hz})$ are shown in Figure 9.13.

Figure 9.13 shows that both legs of the steel frame in the ODS at the motor speed are moving in phase with no relative displacement between two frames near the gearbox region. Hence it is not expected to cause any stress in the gearbox. The ODSs at both gear mesh frequencies $(f_{LGMF} \ \& \ f_{HGMF})$, however, clearly indicate that both frame legs are moving out of phase at f_{LGMF} and in phase at f_{HGMF} with relatively high displacement near the gearbox. Therefore the ODSs at both gear mesh frequencies $(f_{LGMF} \ \& \ f_{HGMF})$ may induce stress in the gearbox and then leading to the frequent gearbox failure.

9.2.3 EXAMPLE 9.3—BLOWER WITH FREQUENT BEARING FAILURE

It is a case study on an Industrial blower system (Balla and Rao, 2004). The blower is driven by a 40 HP, 1460 RPM motor through a V-belt arrangement at a speed of 1460 rpm. The blower shaft is supported by two anti-friction bearings. The blower has 16 blades. The schematic diagram of the blower is shown in Figure 9.14. The blower worked well for the 15 years and then the frequent failure of both bearings started due to high machine vibration. The problem of the frequent bearings failure is identified and solved by the ODS analysis (Balla and Rao, 2004, Sinha and Balla, 2006), is briefly discussed here.

The measured vibration spectra at the bearings B1 and B2 are shown in Figure 9.15, where the bearing B1 is showing dominant belt speed (12.5 Hz) but the dominant vibration peak at the blower speed (17.5 Hz) and the twice of the blade passing frequency (2× BPF) at the bearing B2. The sideband cluster of frequencies at

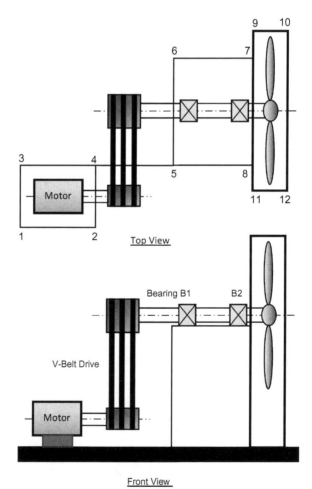

FIGURE 9.14 Schematic of a motor-blowerunit with measurement locations 1–12, B1 and B2.

Operational Deflection Shape (ODS)

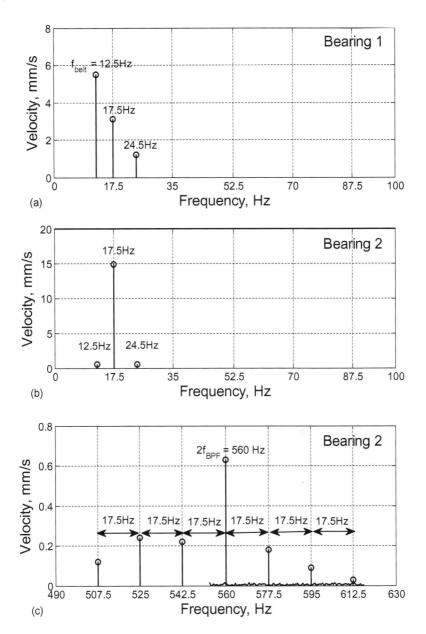

FIGURE 9.15 Measured vibration acceleration spectrum plots, (a) Bearing 1, (b) Bearing 2, and (c) Bearing 2 but in frequency range of 490–630 Hz.

the space of 17.5 Hz is also seen around the frequency (2× BPF). The ODS analysis by Balla et al. (2004) is also done to understand the reason for the frequent failure of both bearings. The ODS at the blower speed in Figure 9.16 shows that the out of phase movement of the bearing pedestals with respect to the blower casing and the motor pedestal. This motion must be loading the bearings and leading to the frequent

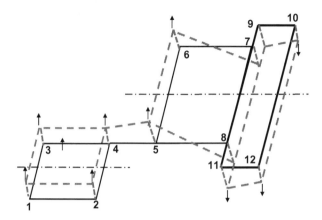

FIGURE 9.16 ODS of the supporting frame of at the motor-blower unit at the blower speed.

failures of bearings B1 and B2. Balla et al. (2004) gave the following possible diagnosis based on the vibration measurements and ODS analysis;

1. Possible looseness in the blade anchoring to shaft due to presence of the 2× BPF frequency component.
2. Possible angular misalignment between the driven pulley and the blower shaft due to appearance of the belt frequency 12.5 Hz.

Both these likely faults are shown in Figure 9.17. Both are rectified to solve the frequent failure of the bearings.

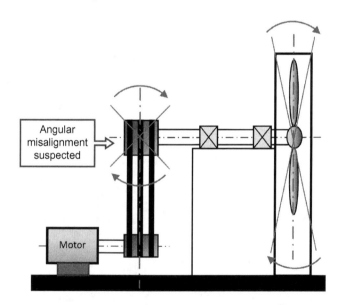

FIGURE 9.17 Identified defects—looseness in the blower blade assembly and pulley wheel alignment.

Operational Deflection Shape (ODS)

9.3 SUMMARY

The step-by-step procedure for the ODS analysis at a frequency is explained through a simple machine. The ODS concept is then demonstrated through a few industrial examples to understand the machine dynamics and also to solve the machine vibration problems. Few more case can be found in the reference (Sinha and Balla, 2006).

REFERENCES

Balla, C.B.N.S., A.R. Rao, 2004. Diagnostics of Exhaust Blower, *Procceeding of 3rd International Conference on Vibration Engineering and Technology of Machinery (Vetomac-3)*, New Delhi, India, December 2004.

Elbhbah, K., J.K. Sinha, W. Hahn, 2018. Solving Gearbox Frequent Failure Problem, *Proceedings of the 3rd income*, Portugal.

Friswell, M.I., J.E. Mottershead, 1999. *Finite Element Model Updating in Structural Dynamics*, Dordrecht, Kluwer Academic Publishers.

Sinha, J.K., C.B.N.S. Balla, 2006. Vibration-based Diagnosis for Ageing Management of Rotating Machinery: A Summary of Cases. *Insight*, 48(8), 481–485.

Sinha, J.K., W. Hahn, K. Elbhbah, G. Tasker, I. Ullah, 2012. Vibration Investigation for Low Pressure Turbine Last Stage Blade Failure in Steam Turbines of a Power Plant. *Proceedings of the ASME TURBO EXPO Conference*, June 11–15, 2012, Copenhagen, Denmark.

10 Shaft Torsional Vibration Measurement

10.1 MEASUREMENT APPROACH

It is difficult to do the direct measurement of the torsional vibration. It requires typical arrangement to do the measurements and then signal processing of the measured data to extract the torsion vibration signal. The torsional vibration is not commonly used in the vibration-based condition monitoring (VCM). However, it is observed to be useful tool to find the defects in the rotating and reciprocating machines.

Following are the simple measurement techniques that can be used to find the shaft torsional vibration. The schematic view of these measurements is shown in Figure 10.1.

1. This approach is making use of any gearwheel or toothed fly wheel on the shaft. Measure the gap of the wheel through the proximity probe (or magnetic pick up, MPU sensor) during the shaft rotation is then required. This concept is similar to the tacho measurement as discussed in Section 4.2.4 of Chapter 4. Here the gear wheel has a number of equally spaced teeth instead of a single keyway or a reflective small width tape.
2. Alternatively an equally spaced black-and-white zebra reflective strip wrapped around the shaft can be used if there is no gearwheel on the shaft. Then measure the zebra strip through the laser sensor during the shaft rotation similar to the laser tacho sensor and its measurement in Section 4.2.4 of Chapter 4.
3. Or a commercially available sensor known as "Encoder" can be used. The concept is exactly same as the option (1) (i.e., "a gearwheel with a sensor"). It consists of a very precisely machined light weight and small gearwheel with a gap sensor. This can be attached to one end of the shaft. The encoders available commercially have a wide range of number of teeth on their gearwheel. Hence depending on the requirements it can be selected. The encoder with higher number of teeth is likely to have better resolution in the shaft torsional vibration measurements.

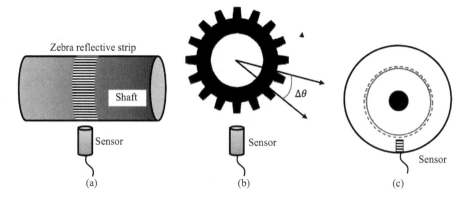

FIGURE 10.1 Simple schematic of different measurement techniques (a) reflective zebra strip on the shaft with a sensor (b) gearwheel mounted on the shaft with a sensor (c) encoder.

10.2 EXTRACTION OF TORSIONAL VIBRATION SIGNAL

Irrespective of any measurement schemes are used, the sensor simply measures the gap between each tooth on the gearwheel during its rotation which results into the pulse train of the measured gap voltage with time as shown in Figure 10.2. There are two possible approaches to extract the shaft torsional vibration from the measured pulse train.

10.2.1 Time Domain Zero-Crossing Approach

The encoder is generally used to measure the shaft speed. The shaft instantaneous speed signal can be extracted based on time-intervals between successive pulses in the encoder raw signal. This instantaneous speed is known as the instantaneous angular speed (IAS) of the shaft. The extraction of instantaneous angular speed (IAS) signal from the measured raw pulse train in Figure 10.2 is discussed in the following steps.

Step 1: Each pulse represents a tooth in the gearwheel, hence the difference of times, $t_2 - t_1$, $t_3 - t_2$, ..., $t_{n+1} - t_n$, etc. represents the time interval required to cross the first, second, ..., nth tooth respectively which is written as

$$\Delta t_n = t_{n+1} - t_n \tag{10.1}$$

Let's assume that the instantaneous shaft time between nth and (n+1)th tooth is measured at the time, t_{ins_n} as

$$t_{ins\_n} = \frac{t_n + t_{n+1}}{2} \tag{10.2}$$

where $n = 1, 2, 3, 4, 5,...$ represents the tooth number and N is the number of teeth in the gearwheel (or number of equally spaced zebra bars).

Shaft Torsional Vibration Measurement

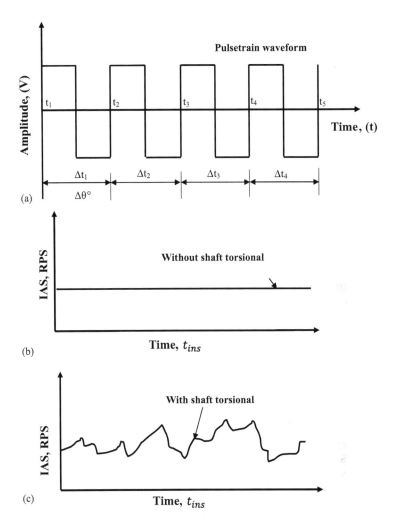

FIGURE 10.2 Typical pulse train waveform from gearwheel (a) and extracted IAS signals (b) without shaft torsional vibration, (c) with shaft torsional vibration.

Step 2: The angular displacement $\Delta\theta$ is the angle between two equally spaced teeth in the gearwheel (see Figure 10.1), which can be written as

$$\Delta\theta = 360°/N = \frac{2\pi}{N} \text{ radian} \tag{10.3}$$

Step 3: The instantaneous angular speed (IAS), $f_{i,n}$ at time, t_{ins_n} in terms of rotation per second (RPS) can be calculated as

$$f_{i,n} = \frac{\Delta\theta}{(\Delta t_n . 2\pi)} \tag{10.4}$$

Hence the extracted IAS is nothing but the representing the shaft torsional vibration. The extracted IAS signals from the measured pulse train are also shown in Figure 10.2. Figure 10.2b represents when a shaft is rotating at a constant speed and no shaft torsional vibration. It is expected to be the ideal case when there is no shaft torsional vibration, that is, each tooth is passing through sensor at an equal time interval, $\Delta t_1 = \Delta t_2 = \Delta t_3 = ... = \Delta t_N$.

However if there is presence of torsional vibration or/and fluctuation of rotating speed then these time intervals, Δt_n, may not be same. Hence the extracted torsional vibration is likely to different than just a straight line. A typical IAS signal is shown Figure 10.2c.

The time vectors, t_{ins_n} (corresponding IAS, $f_{i,n}$) may not be at the equally spaced, hence the data need to be re-sampled or interpolate at a fixed time spacing, dt, that is, at a sampling frequency, $f_s = 1/dt$ before further signal processing.

10.2.2 Demodulation Approach

The spectrum analysis of the measured raw pulse train signal may be similar to one shown in Figure 10.3. The spectrum may contain peaks at 1×, 2×, etc. related to the shaft speed and a dominate peak at a gearwheel frequency, $f_{gearwheel} = N f_{rotor}$, where N is the number of teeth on gearwheel and f_{rotor} is the speed of the rotor.

The sideband frequencies at $f_{gearwheel}$ in Figure 10.3 indicate the modulation of the shaft torsional vibration. This means that the gearwheel frequency, $f_{gearwheel}$ is the carrier frequency that modulates the shaft torsional vibration frequencies. Hence the demodulation of the measured pulse train signal is required to extract the shaft torsional vibration signal.

As explained earlier in Section 7.6.3 of Chapter 7 that it is better to remove rotor related frequencies from the measured raw signal before demodulation. The same process must be followed here as well. Hence a high-pass filtering should be done to remove the rotor related frequencies and then demodulation at the carrier frequency, $f_c = f_{gearwheel}$ or envelope analysis to be carried out on the filtered (remaining) signal to get the shaft torsional vibration signal.

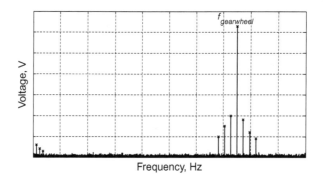

FIGURE 10.3 Spectrum of a measured pulse train waveform signal indicating modulation at the carrier frequency, $f_c = f_{gearwheel}$.

10.3 EXPERIMENTAL EXAMPLES

10.3.1 Example 10.1—Blade Vibration

A part of the study by (Gubran and Sinha, 2014) is discussed here as an example. The photograph of an experimental rig, a single blade and the close view of a bladed disc in the rig are shown in Figure 10.4. The schematic of rig with the vibration instruments is also shown in Figure 10.5 (Gubran and Sinha, 2014). The encoder is mounted at the end of the shaft (Figures 10.4a and 10.5). The encoder used for this measurement has 360 teeth on the gearwheel so that angle ($\Delta\theta$) between two teeth is 1° and generates 360 pulses per shaft rotation.

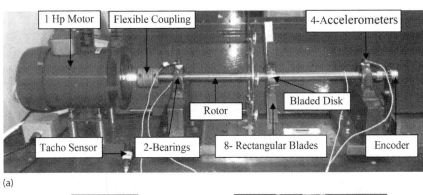

FIGURE 10.4 Photographs of (a) an experimental rig, (b) a single blade, and (c) bladed disc.

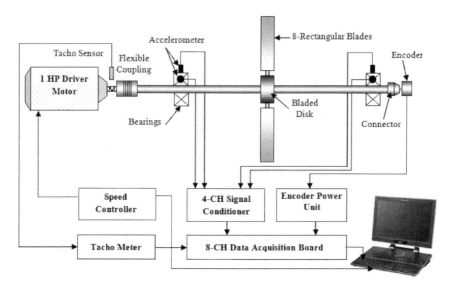

FIGURE 10.5 Schematic of the test rig with instrumentations.

TABLE 10.1
Experimentally Identified Blades 1st Natural Frequency

Blade No.	Blade 1st Natural Frequencies [Hz]
1	123.75
2	126.25
3	125.00
4	127.50
5	128.75
6	123.75
7	125.00
8	125.00
Mean Frequency (±deviation)	125.625 (±3.125)

The bladed disc in the rig contains nearly identical eight blades. The experimentally identified first natural frequency for each blade through *in-situ* modal tests by Gubran and Sinha (2014) is listed in Table 10.1. A small scatter in the blade natural frequencies represents the blade mistuning effect.

The vibration in the blades during machine operation is likely to induce the shaft torsional vibration. A number of vibration measurements were carried out by Gubran and Sinha (2014) during the machine run-ups. Typical machine run-up speed profile and the measured raw pulse train from the encoder are shown in Figures 10.6 and 10.7, respectively. The time domain zero-crossing approach is used to extract the shaft torsional vibration from the measured pulse train waveform and then the STFT is used to analyze torsional vibration.

Shaft Torsional Vibration Measurement

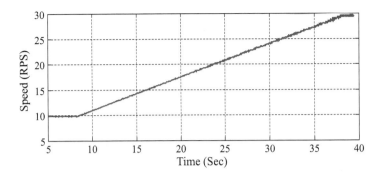

FIGURE 10.6 A typical rotor speed profile for the rig during run-up.

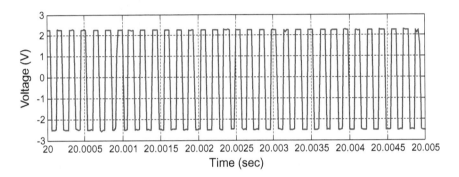

FIGURE 10.7 A typical measured encoder signal.

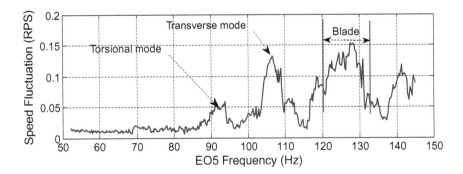

FIGURE 10.8 A typical IAS response at EO5 for the healthy blades.

The 5× of machine run-up speed (i.e., 5th Engine Order, EO5) from the STFT calculation is shown in Figure 10.8. This clearly shows the shaft torsional mode, shaft transverse mode and the banded blade modes due to scatter in the blade natural frequencies. Hence the blade vibration is clearly detected here through the shaft torsional vibration. The difference in the torsional vibration behavior between the

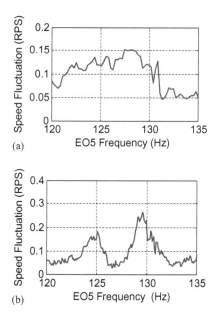

FIGURE 10.9 Typical IAS responses at EO5, (a) healthy blades, (b) crack in two blades.

healthy and crack blades in the blade disc is also shown in Figure 10.9 through this experimental rig. This concept can be used for the blade health monitoring for any rotating machines. Further details are given in the reference (Gubran and Sinha, 2014).

10.3.2 Example 10.2—A Diesel Engine

It is an example of a 4-stroke 16-cylinders diesel engine (Charles et al., 2009). It is a vee form engine consists of A and B banks, and each bank with eight inline cylinders. The power of the engine is 8400 kW and its speed is 740 RPM. Charles et al. (2009) have used the diesel engine flywheel having equally spaced 152 teeth for the crackshaft torsional vibration measurements. The demodulation approach is then used by Charles et al. (2009) to extract the crank shaft torsional vibration from the measured raw pulse train waveform obtained from the flywheel.

Any defect in the combustion process in reciprocating machines is likely to impact the torsional vibration of the crack shaft. The extracted IAS waveforms by Charles et al. (2009) for the healthy and two misfired conditions are shown in Figure 10.10. The IAS waveforms are definitely showing deviations but difficult to do any diagnosis based on the IAS time waveforms only. The spectra of these IAS signals are also shown in Figure 10.11 where the firing frequency peak at around 100 Hz is clear seen for the healthy condition but no peak at firing frequency at around 100 Hz for misfired cases.

The firing frequency for the diesel engine is calculated as

$$f_f = \frac{1}{T_f} \tag{10.5}$$

Shaft Torsional Vibration Measurement 217

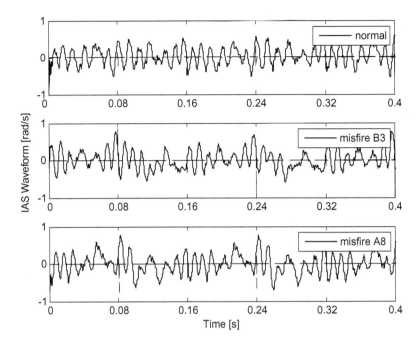

FIGURE 10.10 Measured IAS signals of a 16-cylinder diesel engine with different fault conditions.

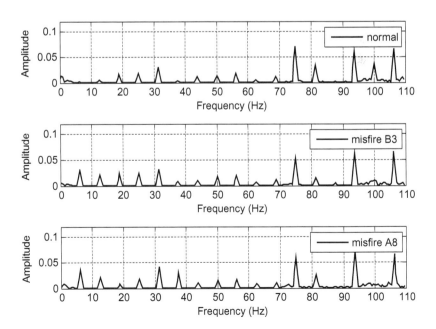

FIGURE 10.11 Spectra of the IAS signals in Figure 10.10.

where T_f is the period of time between the consecutive cylinders firing, which is given by

$$T_f = \frac{T_{cs}}{N_c} \qquad (10.6)$$

where T_{cs} is the time for one complete engine cycle. For four strokes, it is equal to two complete rotations of the crack shaft (i.e., 720°). N_c is number of cylinders in the engine. Here $T_{cs} = 2 \times \frac{60}{740} = 0.1621$s, $N_c = 16$ so $T_f = \frac{T_{cs}}{N_c} = 0.0101$ s and $f_f = \frac{1}{T_f} = 99.01$ Hz

10.4 SUMMARY

The concept of the pulse train measurement using the gearwheel or reflective black and white zebra strip is discussed in the chapter. A couple of methods used to extract the shaft torsional vibration from the measured pulse train are also discussed. A couple of experimental examples are also presented in the chapter to demonstrate the measurements and data process for the shaft torsional vibration and their usefulness.

REFERENCES

Charles, P., J.K. Sinha, F. Gu, L. Lidstone, A.D. Ball, 2009. Detecting the Crankshaft Torsional Vibration of Diesel Engines for Combustion Related Diagnosis. *Journal of Sound and Vibration*, 321(3), pp. 1171–1185.

Gubran, A., J.K. Sinha, 2014. Shaft Instantaneous Angular Speed (IAS) for Blade Vibration in Rotating Machine. *Mechanical Systems and Signal Processing*, 44(1–2), pp. 47–59.

11 Selection of Transducers and Data Analyzer for a Machine

11.1 INTRODUCTION

The selection of instrumentations for vibration measurements, signal collection and data analyses capabilities of the analyzer are most important consideration in the vibration-based condition monitoring. This chapter discusses about these requirements through an example of a simple machine.

11.2 CALCULATION OF MACHINE FAULTS FREQUENCIES

The schematic of a horizontal centrifugal pump driven by a motor is shown in Figure 11.1. Following are the machine details (mostly assumed values for demonstration).

540 kW 3 phase Motor,

Motor Speed = 3000 RPM $(50\,\text{Hz})$,

Discharge rate = 25000 LPM

7 impeller blades in the pump,

5 bearings $(1$ for centrifugal pump, 2 bearings for motor—drive end & non-drive end and 2 roller bearings on either side of flywheel$)$

The assumed roller bearing specifications are listed in Table 11.1.

Based on the above information the calculated frequencies related to different faults for this machine are listed in Table 11.2.

219

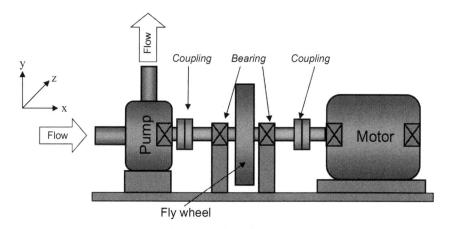

FIGURE 11.1 Schematic of a pump assembly.

TABLE 11.1
Bearing Details

Pitch circle diameter (d_p)	195 mm
Diameter of roller (d_b)	15 mm
Contact angle of the roller (β)	0
Number of rollers (n)	30
Inner (bore) diameter	160 mm

TABLE 11.2
Frequencies of Different Faults

Faults	Equation for Fault Frequency	Frequency, Hz
Rotor Faults		
Unbalance, misalignment, crack, rub, bend, etc	Up to 10× is likely to cover all rotor related faults	500 Hz
Motor Faults		
General electric problem	Harmonics of rotating speed frequency and harmonics of line frequencies—up to 10×	500 Hz
Stator defect (short circuits)	Even harmonics of line frequency 50 Hz. Consider up to 10× f	500 Hz
Rotor defect (broken bar)	Side bands due to slip around the machine speed frequency and its harmonics—up to 5×	250 Hz
Other Machine Faults		
Impeller blades	Up to 5 x BPF, BPF = 7× = 350 Hz	1750 Hz
Mechanical looseness (including loose assembly of bearings)	Several harmonics and sub-harmonics of the machine rotating speed frequency; consider up to 10×	500 Hz

(Continued)

Selection of Transducers and Data Analyzer for a Machine

TABLE 11.2 (*Continued*)
Frequencies of Different Faults

Faults	Equation for Fault Frequency	Frequency, Hz
Bearing Faults—Equations from Chapter 7 are used		
Inner race defect	$f_i = \dfrac{n}{2} f_r \left(1 + \dfrac{d_b}{d_p} \cos\beta\right)$	807.69 Hz
Outer race defect	$f_o = \dfrac{n}{2} f_r \left(1 - \dfrac{d_b}{d_p} \cos\beta\right)$	692.30 Hz
Roller defect	$f_b = \dfrac{d_p}{d_b} f_r \left(1 - \left(\dfrac{d_b}{d_p}\right)^2 \cos\beta^2\right)$	464.15 Hz
Cage defect	$f_{cage} = \dfrac{f_0}{n}$	23.08 Hz
Bearing housing and assembly resonance frequency. Bearing defect frequency often get modulated with this resonance frequency at the early stage of bearing defect.	Better to conduct *in-situ* modal tests on the bearing housing to find out the resonance frequency—assume here up to 5 kHz	5 kHz

11.3 SELECTION OF ACCELEROMETER

It is known from the principle of the accelerometer that the accelerometer can measure the vibration accurately up to 20% of the accelerometer natural frequency. Here the maximum frequency of 5 kHz is enough to cover all defect frequencies; therefore, the accelerometer must measure the vibration accurately up to 5 kHz. Therefore the accelerometer with natural frequency of 25 kHz or more is going to be appropriate for this machine. It is also important to consider the following before selection of the accelerometer.

1. Consider the accelerometer mounting arrangement. It is likely that the natural frequency of the accelerometer may not alter if the stud mounting is used. But for any other mounting arrangements, the mounted natural frequency is likely to be lower than design value and hence the lower frequency ranges for the linear and accurate vibration measurement. Therefore it is better to have some margin while selecting the accelerometer to keep the linear and accurate vibration measurement up to 5 kHz even if you choose different mounting arrangement for the accelerometer.
2. Accuracy (both amplitude and phase) in vibration measurement.
3. Working temperature—the selected accelerometer must withstand the working temperature environment.

11.4 ANALYSIS PARAMETERS

Following data analyses are required for the vibration monitoring and fault diagnosis on the measured vibration data from the machine.

11.4.1 Time Domain Analyses

At least the following basic steps are essential for the VCM.

1. Filtering as appropriate to condition the measured signals, e.g., removal of low frequencies noise using the high pass filter, high pass or band pass filter to restrict the data into the required frequency range.
2. Estimation of the crest factor (CF) and/or kurtosis (Ku) for the measured vibration signal up to 5 kHz to understand the nature of vibration signal. This may be useful for the bearing health monitoring.
3. Estimation of vibration RMS value in the velocity domain up to 5 kHz so that it can be compared with the appropriate ISO code for the vibration severity limits.

11.4.2 Frequency Domain Analyses

Once again the following basic steps are important.

1. Spectrum analysis with required frequency resolution in both acceleration and velocity domains.
2. Velocity spectrum for the rotor related defect detection.
3. Acceleration spectrum to observe if there is any hump in the high frequency range related to bearing housing and assembly resonance frequency.
4. Phase analysis at 1×, 2×, ... etc. using tacho signal if required.

11.4.3 Time-Frequency Analyses

It is also important to do the following analysis if required.

1. Waterfall 3D spectra analysis during the machine run-up and run-down if needed.
2. Envelope analysis of the measured acceleration signal for bearing fault diagnosis. Initially the measured vibration acceleration signal is required to filter using either high pass filter to remove rotor related frequencies or band pass filter having only modulated signal of the bearing resonance with bearing defects frequencies (if any). Extract the envelope signal from the filtered signal and then estimate the spectrum of the envelope signal to do the bearing health diagnosis.

11.5 FEATURES REQUIRED IN THE DATA ANALYZER

Data analyzers available commercially are generally a combination of hardware and software. Following specifications and analyses features should be required.

Selection of Transducers and Data Analyzer for a Machine 223

11.5.1 Specifications

1. Number of input channels: This depends upon the requirement but it is better to have at least two input channels for any portable system. This two-channel analyzer becomes handy if anyone wants to perform modal tests and phase analysis using tacho signal. One channel can be used for the instrumented hammer or tacho signal and other for the vibration measurement to perform modal tests or phase analysis, respectively.
2. Analogue-to-digital-conversion (ADC) device: It is better to have 16 bit or more for the ADC during data collection in the vibration measurement process.
3. Sampling frequency: Analyzer must support the required sampling frequency simultaneously for all input channels. For the example considered the analyzer must acquire the data the sampling frequency of $fs = 2.56 \times fu = 2.56 \times 5$ kHz or more.
4. Input voltage per channel: Many commercially available sensors can generate the maximum voltage of ± 5 V. So maximum input voltage for each channel should be at least ± 5 V or more (Vmax). The system should have flexible options for selecting different input voltage range between 0 and $\pm$Vmax. This option can improve the measured signal quality (signal to noise ratio) if the selection of input voltage range per input channel is close to the signal voltage level from the sensor.
5. Trigger option: At least one channel of the analyzer must have a trigger option to start the data collection-based pre-set trigger level. This will help modal tests and phase analysis using the hammer and tacho sensor, respectively.
6. Anti-aliasing filter and other filters: The system must have an inbuilt anti-aliasing filter at half of the sampling frequency to avoid aliasing effect in the data collection. This should also have different filters to meet the other required data processing for the machine vibration moniotoring.
7. Online and offline data analysis capabilities: An analyzer should be able to do the pre-set sequence of analysis during the measured process. In addition, the analyzer system should have temporary storage to store measured raw data so that these data can be analyzed offline to further investigate the measured data if needed.

11.5.2 Data Analysis Capabilities

1. Filtering—LP, HP and BP filters
2. Data integration and differentiation in both time and frequency domains
3. RMS, peak, peak-to-peak, CF, Ku, etc. estimation
4. Spectrum analysis using selectable frequency resolution, df.
5. CPD, FRF and coherence analysis between two signals
6. Envelope analysis

11.5.3 Data Trending and Storage

It is preferred that the system should have at least the following capabilities.

1. Storage facility to store the measured spectra or raw data for six or more months depending upon requirements.
2. Data trending and storage features for RMS velocity, Acceleration CF/Ku, the values of 1×, 2×, 3× and other frequencies required for the machine vibration monitoring and their phases depending upon requirements.

11.6 SUMMARY

The chapter only brings out the basic requirements and guidance for the instrumentation selection for the machine vibration monitoring. Similar approach should be considered for any other machines used in industries.

12 Future Trend in VCM

12.1 INTRODUCTION

Vibration-based condition monitoring (VCM) of rotating machines is a well-accepted tool for the condition-based maintenance (CBM) activity in industries (Sinha, 2015). The VCM generally helps to predict any developing fault(s) or any deviation in the machine dynamics behavior at early stage so that the maintenance or remedial action can be performed before any catastrophic failure. If this approach is applied efficiently, then the machine downtime, maintenance overhead can be reduced significantly with enhanced the plant safety.

The data collection and signal processing required in the VCM is now much easier due to a number of technology advancements in instrumentations and signal processing capabilities over—two to three decades. However the fault detection process is not straightforward in many cases even with these advancements in the technologies. It is because the manual approach is often used to do the fault detection process. It is also observed that it is difficult to use the historical vibration data from a machine to apply directly to aid the fault detection process in many identical machines. But it is very common in a plant to use many numbers of identical machines to meet the production requirements. The application of the VCM-based fault detection process individually for each identical machine at a plant site or many plant sites is definitely a time consuming process. There are following possible limitations in the current VCM practices.

Limitation 1: Figure 12.1 demonstrates the typical current VCM approach for the detection of any fault(s). The vibration measurements are generally carried out at each bearing pedestal in the machine in three mutually perpendicular directions (e.g., vertical, lateral and axial directions) for a horizontal machine as per the ISO code. Overall vibration value (RMS vibration velocity) at is then compared with the ISO severity limits. The presence of harmonic peaks (1×, 2×, 3×,...) and/or sub-harmonic peaks related to the machine RPM in the vibration spectrum/spectra and their amplitudes trending is then used to identify the presence of different faults.

Many industries are also using a slightly more involved signal processing tools in their VCM system for their rotating machines, which are listed in Figure 12.2. In this approach, the measured vibration data during the machine normal or/and transient operation are generally analyzed in a number of different formats to identify many well-recognized machine faults/defects.

However the fault identification process is not very straightforward, and often data intensive and complex process due to large volume of vibration

226 Industrial Approaches in Vibration-Based Condition Monitoring

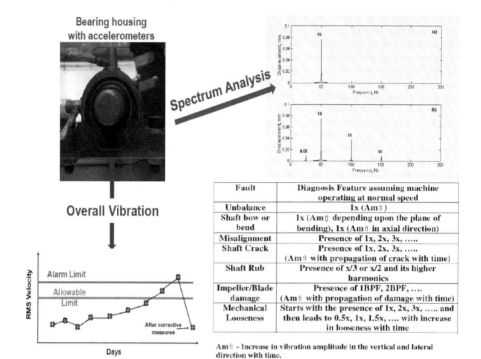

FIGURE 12.1 Simple VCM approach.

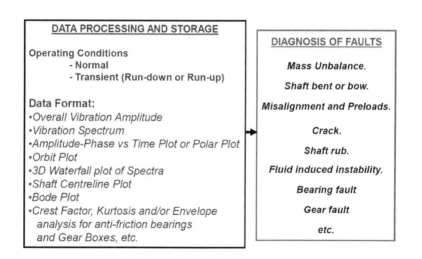

FIGURE 12.2 Different data processing used in the VCM approach.

Future Trend in VCM

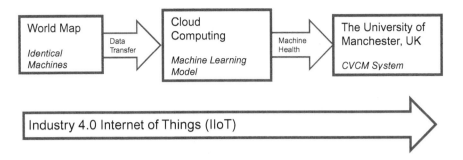

FIGURE 12.3 Typical abstract representation of the proposed CVCM system.

data measured at all bearing locations in three directions from a machine. The fault identification is observed to be subjective process and requires experience and engineering judgments.

Limitation 2: There are several but exactly identical machines that are used around the globe, but it is often difficult to apply the historical vibration data from a machine to aid the fault detection process in many identical machines. It is due to the fact that the structural foundation for each machine installation may be different that generally results in different dynamic behavior.

12.1.1 Future IIoT-Based CVCM Approach

Hence a generic approach is required to make the fault detection process easy. Sinha and his co-authors are extensively involved in the development of such a generic approach using machine learning (ML) approaches. The concept of the Industrial Internet of Things (IoT) 4.0 is gaining momentum in the plant maintenance. It is now feasible to develop the centralized VCM (CVCM) system using the machine learning (ML) approach together with the IIoT for the early detection of faults in any machines. A typical abstract of the possible CVCM system is shown in Figure 12.3. The ML model in the cloud computing environment will have access to the vibration data from identical machines across the globe to do the CVCM.

This chapter presents the future feasibility and possibility of the CVCM systems for the identical machines around the globe for the existing old and new plants.

12.2 APPROACH 1: SUITABLE FOR EXISTING OLD PLANTS

The horizontal centrifugal pump shown in Figure 12.4 is considered here to explain the concept. There are many of the exact same pumps that can be in used in a plant or in many plants across the globe for decades. It is also possible that many pumps are already monitored by the VCM approach over last couple of decades. So there is possibility of a data bank of the processed vibration data such as overall vibration velocity (V_{RMS}), crest factor and/or kurtosis, and the vibration values at 1×, 2×, 3×,

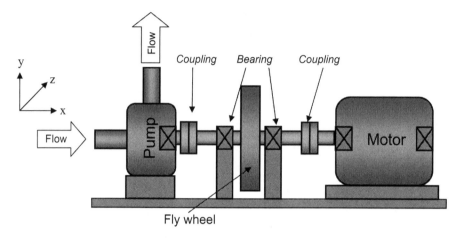

FIGURE 12.4 A horizontal centrifugal pump.

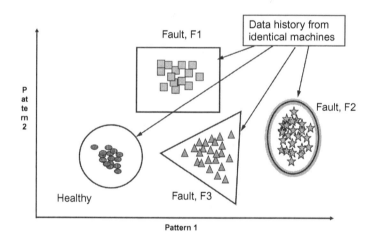

FIGURE 12.5 Faults classification.

etc. at each bearings together with machine health conditions. It is good to compile all the data together with the machine health conditions to generate a data bank.

The big data analysis and artificial intelligence (AI) can then be used to develop the ML model using the data bank. The developed ML model is trained such that the data of different faults must be classified in different baskets as shown in Figure 12.5. Once the model is ready it can then predict the condition of the machine health based on the new data set from the vibration measurements. The studies by Nembhard and Sinha (2015), Nembhard, Sinha et al. (2015) and Luwei et al. (2018) are similar to this approach.

12.3 APPROACH 2: SUITABLE FOR NEW PLANTS

Most of the machines are having more than a bearing and accepted practice to do the vibration measurements in three orthogonal directions at each bearing as per the ISO 20816 code. Hence if any machine has 10 bearings then the total number of measurements is likely to be 30 or more if the measurements are done using both accelerometers and proximity probes. Therefore 30 or more vibration spectra are to be analyzed to identify machine health condition. This makes the process very complex.

Sinha (2008) initially proposed the data fusion in the frequency domain to construct a single composite spectrum and higher order spectra for a machine. This proposed approach has reduced the number of spectra from the several measurement locations to a single vibration spectrum to easy the diagnosis process. The typical concept is shown in Figure 12.6. This approach is likely to reduce the size of data transfer to the ML model through IIoT and hence the significantly reduced data bank size as well. This approach can be applied to new but identical machines from the healthy condition itself.

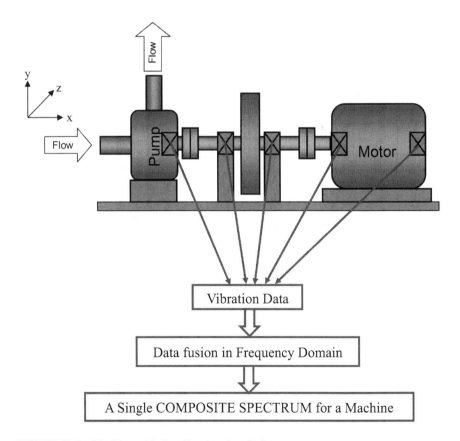

FIGURE 12.6 Machine with the vibration data fusion concept.

230   Industrial Approaches in Vibration-Based Condition Monitoring

Elbhbah and Sinha (2012a, 2012b, 2013) have then applied this composite spectrum approach to an experimental rig having four bearings. Elbhbah and Sinha (2012a, 2013) have also introduced the coherence (Sinha, 2105) between the measured vibration responses at different bearings to reduce the noise in the measured vibration signals. The coherent composite spectrum found to be more robust in the fault identification process (Elbhbah and Sinha, 2013). This concept has then been further extended to the coherent higher order spectrum that has the advantage of combining different harmonics components (both phases and amplitudes) within a spectrum (Sinha and Elbhbah, 2013).

Earlier data fusion approach to construct the coherent composite spectrum has also been improved further so that both the amplitudes and phases information between multiple signals can be used in more effective way (Yunusa-Kaltungo et al., 2014). The improved coherent composite spectrum is named as the *poly* coherent composite spectrum (*pCCS*). The pCCS and its higher order spectra (Yunusa-Kaltungo et al., 2015, Yunusa-Kaltungo and Sinha, 2017) have further been tested in the identification of the defects in rotating machines.

Figure 12.6 also represents the concept of data fusion of the measured vibration at all bearings into a single composite spectrum (*pCCS*). The computation of *pCCS* is given in Equation (12.1) (Yunusa-Kaltungo et al., 2014).

$$S_{pCCS}(f_k)$$

$$= \left(\frac{\sum_{r=1}^{n_s} X_1^r(f_k)\gamma_{12}^2 X_2^r(f_k)\gamma_{23}^2 X_3^r(f_k)\gamma_{34}^2 X_4^r(f_k)\dots X_{(b-1)}^r(f_k)\gamma_{(b-1)b}^2 X_b^r(f_k)}{n_s} \right)^{(1/b)}$$

$$(12.1)$$

where $X_1^r(f_k)$, $X_2^r(f_k)$, $X_3^r(f_k)$, $X_4^r(f_k)$,....., $X_{(b-1)}^r(f_k)$ and $X_b^r(f_k)$ are the Fourier transformation (FT) at frequency f_k of the *r*th segment of the vibration responses at bearings 1, 2, 3, 4,....., (b–1) and b. γ_{12}^2, γ_{23}^2, γ_{34}^2,....., $\gamma_{(b-1)b}^2$ are the coherences between bearings 1-2, 2-3, 3-4, …, (b–1)–b, respectively. n_s is the number of data segments of the measured responses. The FT and coherence are computed using Equations (5.10) and (5.19), respectively.

Initially it is better to develop the self-learning ML model using vibration data from the *pCCS* components related to 1×, 2×, 3×, etc. from new machines in healthy condition and then keep enriching the ML model with more data from the identical machines with time of operation. If there is any fault developed during machine operation then the new data may not appear in the healthy basket. Find out the exact nature of fault and then define another basket in the self-learning ML model for this fault.

12.4 SUMMARY

The suggested approaches are expected to lift the limitations of experience of maintenance professionals, engineers and practitioners involved in the VCM and fault detection process. The CVCM can be managed centrally irrespective of machines installation locations. This seems to be a realistic and feasible approach in future.

REFERENCES

Elbhbah, K., J.K. Sinha, 2012a. Comparison of Vibration Based Coherent Composite Spectrum with Non-coherent Composite Spectrum for Machine Condition Monitoring', 15th *International Conference on Experimental Mechanics*, Paper Ref. 2940, Faculty of Engineering, University of Porto, Portugal, July 22–27.

Elbhbah, K., J.K. Sinha, 2012b. A Composite Vibration Spectrum for a Machine for Vibration Based Condition Monitoring, *Proceedings of the ASME 2012 International Design Engineering Technical Conferences & Computers and Information in Engineering Conference* DETC2012/70052, August 12–15, Chicago, IL.

Elbhbah, K., J.K. Sinha, 2013. Vibration-based Condition Monitoring of Rotating Machines using a Machine Composite Spectrum. *Journal of Sound and Vibration*, 332(11), pp. 2831–2845.

Luwei, K.C., J.K. Sinha, A. Yunusa-Kaltungo, K. Elbhbah, 2018. Data Fusion of Acceleration and Velocity Features (dFAVF) Approach for Fault Diagnosis in Rotating Machines, *VETOMAC Conference*, Lisbon, Portugal, September 10–13, 2018.

Nembhard, A., J.K. Sinha, 2015. Unified Multi-speed Analysis (UMA) for the Condition Monitoring of Aero-Engines. *Mechanical Systems and Signal Processing*, 64–65, pp. 84–99.

Nembhard, A., J.K. Sinha, A. Yunusa-Kaltungo, 2015. Development of a Generic Rotating Machinery Fault Diagnosis Approach Insensitive to Machine Speed and Support Type. *Journal of Sound and Vibration*, 337, pp. 321–341.

Sinha, J.K., 2008. Bi-spectrum of a Composite Coherent Cross-Spectrum for Faults Detection in Rotating Machines', *IMechE* 9th *International Conference on Vibrations in Rotating Machinery*, University of Exeter, UK. September 8–10, 2008.

Sinha, J.K., 2015. *Vibration Analysis, Instruments, and Signal Processing*, Boca Raton, FL: CRC Press/Taylor & Francis Group.

Sinha, J.K., K. Elbhbah, 2013. A Future Possibility of Vibration Based Condition Monitoring. *Mechanical Systems and Signal Processing*, 34(1–2), pp. 231–240.

Yunusa-Kaltungo, A., J.K. Sinha, 2017. Effective Vibration-Based Condition Monitoring (eVCM) of Rotating Machines. *Journal of Quality in Maintenance Engineering*, 23(3), pp. 279–296.

Yunusa-Kaltungo, A., J.K. Sinha, K. Elbhbah, 2014. An Improved Data Fusion Technique for Faults Diagnosis in Rotating Machines. *Measurement*, 58, pp. 27–32.

Yunusa-Kaltungo, A., J.K. Sinha, A. Nembhard, 2015. A Novel Faults Diagnosis Technique for Enhancing Maintenance and Reliability of Rotating Machines. *Structural Health Monitoring: An International Journal*, 14(6), pp. 604–621.

Index

Note: Page numbers in italic and bold refer to figures and tables, respectively.

A

acceleration sensors, 48–53
acceleration spectrum, *100*, *145*, *205*
   on anti-friction bearing, *138*
   to displacement spectrum, conversion, 99, **99**
   on gearbox, *150*
   for healthy machine, *123*
   response, *164*
   size defects, *123*, *124*
accelerometer, *48*, 48–49, *49*
   mounting, 51, *52*, 53
   in non-dimensional form, *51*
   selection, 221
   specifications, **52**
accuracy, 60
acoustic emission monitoring, 3
ac signal, 116
ADC (analog-to-digital conversion) process, 58, 61, 66
aged machine, 42
aliasing effect, 61, 63–66
AM (amplitude modulation), *109*
amplification factor, 20–21, 32
amplitude
   with DAQ device, *69*
   FRF, 106, *106*, *167*, 168, *174*
   linearity, 51
   phase *versus* time data, 115, *116*
   spectrum, *85*, 94, *162*, *173*
   unbalance force, 32
   vibration, 74–76, *76*, 113, *114*
amplitude modulation (AM), *109*
analog-to-digital conversion (ADC) process, 58, 61, 66
angular displacement, 211
anti-aliasing filter, 61, 63–67, *66*, *67*, 73
anti-friction bearing fault detection, 137–140, *138*, *139*
   CF, 141
   envelope analysis, 142–143
   Ku, 141–142
averaging process, 92–93, *93*, *94*
   key elements, **96**
   overlap in, 94–95

B

band pass (BP) filter, 73, *74*, 222
bath tub concept, 35
   life cycle model, *36*
   modification, *43*
beam acceleration response, *172*
bearing(s), 38, 118
   anti-friction, 38, *137*, 137–140, *138*
   blower with failure, 204–206
   CF/Ku for, *142*
   details, **144**, **220**
   failed drive-end motor, *147*
   fluid, 151–152
   FRF phase plots, *198*
   journal, *151*
   machine vibration on, *30*
   measurements, *39*, *40*
   pedestal, 129, 184, *185*, 199
   resonance response, *139*
   roller, 137, *143*, 143–144
   STFT plot, *102*
   vibration acceleration, *29*, *195*, *205*
   vibration sensors, *56*
Blade Health Monitoring (BHM), 129
Blade Passing Frequency (BPF), 129
Blade Tip Timing (BTT) method, 129
blade vibration, 129, *213*, 213–216, *214*, **214**
Bode plot, 119, *120*
BPF (Blade Passing Frequency), 129
BP (band pass) filter, 73, *74*, 222
Breakdown Maintenance (BM), 4
BTT (Blade Tip Timing) method, 129
bump test, 159

C

calibration chart, 59
cantilever beam, 24, *24*
   modeshapes, *25*
   vibration mode, *26*, *27*
carrier frequency, 109
CBM, *see* Condition-based Maintenance (CBM)
centralized VCM (CVCM) system, *227*, *227*
centrifugal force, 27, 29, *29*
CF (Crest Factor), 78, 141, 222

233

234 Index

clamped-clamped beam, *170*, 170–178, *171*, *174*, **176**, *177*
CM, *see* Condition Monitoring (CM)
coherence
  composite spectrum approach, 230
  computation, 162
  FRF and, *108*
  function plot, 106, *107*
  ordinary, 104–105
  two simulated signals with noise, **105**, 105–106, *106*
complimentary solution, 18–19
Condition-based Maintenance (CBM), *4*, 4–5
  CM leading, *6*
  machines for, *5*
Condition Monitoring (CM), 1, *2*, 36
  leading to CBM, *6*
  techniques, 1, 3
Crest Factor (CF), 78, 141, 222
critically damped system, 15, *16*
cross power spectrum, 103–104
CVCM (centralized VCM) system, 227, *227*

D

D'Alembert Principle, 11
damped SDOF system, *14*
damped system, 13–14
  under, 16–17, *18*
  critical, 15, *16*
  over, 15–16, *16*
damping ratio, 15
  HPP method, **179**
data acquisition (DAQ) device, 45, 56, 67
  ADC using 16-bit, **68**
  collected data for, **70**
  multi-input channels, *58*
  signal amplitudes with, *69*
data analyzer, 222
  capabilities, 223
  data trending and storage, 224
  specifications, 223
data collection steps, *57*, 57–58
data sampling rate, 61
dc signal, 116
Debris analysis, 3
demodulation approach, 212
diesel engine, 216, 218
Digital FT (DFT), 84–85
displacement sensors, 45–46
displacement spectra, *100*
Doppler effect, 47–48
dynamic equilibrium, 11

E

eddy current probe, 45–46
electric motor defects, 129–130

encoder, 209–210, *210*, 213
  signal, *215*
envelope analysis, 222
  signal processing, 109–111, *110*
  steps, *135*, *136*
equation of motion, 11–13
  for free vibration, 14–15
  SDOF, 31
experimental modal analysis, 159, 162–169, *163*, *164*
  clamped-clamped beam, 170–178
  impulsive load using hammer, 159–162
  procedure, 159
experimental rig, *213*
  PSD, 96
  rotating rig-1, 178–180, *179*, *180*, *181*
  rotating rig-2, 181–183, *182*
  STFT analysis, 101, *102*
  with vibration instruments, *28*
  vibration spectrum plots, *146*

F

fan-gearbox-motor (FGM), 148
Fast FT (FFT), 85, *174*
fault detection process, 225
faults types, 125
field rotor balancing, 153–157
filtering process, 222
fluid bearings, 151–152
forced vibration, 18–22, *21*
force sensor, 161
forward calculation procedure, 71, *71*
Fourier transformation (FT), 82
  computation, 84–87, **85**
  frequency resolution, importance, 87–88, *88*
  leakage, 88–89, *89*
  sine wave signal, 83, *83*
  window functions, 89–92, *90*
frequency domain analyses, 222
frequency modulation (FT), *109*
frequency ratio, 22
frequency response function (FRF), 104, 195–196, *198*
  amplitude/real/imaginary/phase, *167*, *174*
  bearing pedestal, *185*
  and coherence plots, *108*
  computation, 162
  measured, *175*
  non-dimensional, *169*
  two simulated signals with noise, **105**, 105–106, *106*
  zoomed view, *169*
FSIV (full scale input voltage), 67–68
FS (full scale) reading, 60
FT, *see* Fourier transformation (FT); frequency modulation (FT)

Index

full scale input voltage (FSIV), 67–68
full scale (FS) reading, 60

G

gearbox
    conditions, **134**
    failure, *202*, 202–203, *203*
    fault, 148–151, *150*
    fault detection, 130–135, *131*, *135*
    single stage, 130
    spectrum plots, *133*
    typical vibration, *132*
    vibration spectra, *134*
gear mesh frequency (GMF), 131,
    149, **150**
gear ratio (GR), 130
gearwheel frequency, 212

H

half-power point (HPP) method, 168–169
high pass (HP) filter, 73, *74*, 212
high pressure (HP) turbine, 198, *200*
horizontal centrifugal pump, 183–186, *184*, *185*,
    186, 227, *228*

I

IAS, *see* instantaneous angular speed (IAS)
impulse-response method, 159, 179
impulsive response, *138*, 140, 142
industrial blower system, *204*, 204–206
industrial centrifugal pump, 101–103,
    102, *103*
industrial IoT 4.0, 227, *227*
inertia force, 11
influence coefficient method, 126, 153
in situ modal tests, 159, 186–188
instantaneous angular speed (IAS),
    210–212
    response at EO5, *215*, *216*
    16-cylinder diesel engine, *217*
    spectra, *217*
    waveforms, 216
instrumented hammer, 159, *160*, *172*
    force sensor in, 161, 170
    impact heads, *161*
    for modal testing, *171*
intermediate pressure (IP) turbines,
    198, *200*
ISO codes, 38–39, 41–42, 122

K

kurtosis (Ku), 79, 141–142, 222

L

laser velocity sensor, 47–48
Lead-Time-to-Maintenance (LTM), 5–8
    estimating, *8*
    for machine, *7*
    parameters, **8**
leak detection monitoring, 3
linearity, 60
Lower LTM (LLTM), 7
low pass (LP) filter, 73, *74*
low pressure (LP) turbines, 198, *200*
LTM, *see* Lead-Time-to-Maintenance (LTM)
lubricant monitoring, 3

M

MAC (modal assurance criteria), 176–177, **178**
machine critical speeds, 119
machine fault detection, 129
    BHM, 129
    BPF, 129
    electric motor defects, 129–130
    mechanical looseness, 129
machine faults
    classification, *228*
    frequencies, 219, **220–221**
    machine vibration, 32, *33*
machine learning (ML) approach, 227–228
machine vibration, 25
    on bearing housing, *30*
    machine faults, 32, *33*
    rotor dynamics, 25–26, *28*, *29*
    under rotor unbalance, *31*
    sinusoidal response, 31, *32*
    unbalance responses, 27, 29–32, *30*
magnification factor, 20
mass unbalance, 125–126
mean LTM (MLTM), 8
mean-time-to-failure (MTTF), 1
mechanical looseness, 129
ML (machine learning) approach, 227–228
modal assurance criteria (MAC), 176–177, **178**
modal testing, 162; *see also* experimental modal
    analysis
modeshapes concept, 23–25, **24**
    cantilever beam, *25*
    for clamped-clamped beam, *177*
    experimental rig-1, *181*
    experimental rig-2, *183*
    extraction, 176–177
    machine, 193, *194*
    measured, *183*
    at natural frequency, 184, *184*
    normalized, **176**
    pump assembly and, *187*
    rotor, *28*

N

natural frequency, 12
noise monitoring, 3
non-destructive test (NDT) techniques, 1, 3
non-dimensional CF, *142*
normal operation condition
    amplitude, 115, *115*
    orbit plot, 116–117, *117*
    polar plot, 115, *115*
    vibration amplitude, 113, *114*
    vibration spectrum, 113, *114*, 115

O

ODS, *see* operational deflection shape (ODS)
        analysis
oil whip/whirl, 151–152, *152*
operational deflection shape (ODS)
        analysis, 36
    at critical speeds, *201*
    gearbox failure, 202–203
    at gear mesh frequency, 203
    industrial blower system, 204–206
    at motor-blower unit, *206*
    motor-gearbox-fan, *203*
    at operating speeds, *202*
    rotating machine, 193, *194*
    simple theoretical concept, 193–198
    steam TG set, 198–201
    values at 1× and 2×, 197, **197**, *199*
orbit plot, 127, 152
    normal operation condition, 116–117, *117*
    at rotor speed, 128, *128*
    transient operation conditions, 118
ordinary coherence, 104–105
oscillation, 11–13, *13*
over damped system, 15–16, *16*

P

PA-Fan, *see* primary air fan (PA-Fan)
particular integration, 19
*p*CCS (*poly* coherent composite
        spectrum), 230
P-F (Potential Failure) curve, 40, *41*
phase distortion, 51
phase *versus* time plot, 115, *115*
piezoelectric accelerometers, 48
Planned Preventive Maintenance (PPM), 4
polar plot analysis, 154–155
poles, representation, 17, *17*
poly coherent composite spectrum
        (*p*CCS), 230

motor defects, features, **130**
motor-gearbox-fan unit, *149*, 202, *202*

power spectral density (PSD), 93
    averaging process, 92–95, *93*, *94*
    experimental rig, 96, *97*
    industrial blower, 96–98, *98*
primary air fan (PA-Fan), *147*, 147–148, *148*
proximity probes, 45–46, *47*, 116
PSD, *see* power spectral density (PSD)
pulse train waveform, 210
    from gearwheel, *211*
    spectrum, *212*
pumps/piping
    assembly, *187*, 220
    horizontal centrifugal, 183–186, *228*
    industrial centrifugal, 101–103, *102*
    schematic layout, *187*
    stool, stiffening, *188*
    vertical centrifugal, 186–188

R

reference signal, 53
Reliability and Maintenance Index (RMI), 6–7
repeatability, 60
reproducibility, 60
resonance, 20
    fluid, 151–152
    frequency, 52, 137, 139
    response built-up, 20, *21*
reverse calculation procedure, 71, *71*
RMI (Reliability and Maintenance Index), 6–7
RMS, *see* root mean square (RMS)
roller bearing defect, *143*, 143–144, *145*
root mean square (RMS), 74–76, 113
    amplitude, vibration, *114*
    time wave forms, *76*
    velocity, 124
    vibration values, **124**
    vibration velocity, 147
rotating machine representation, 121, *122*
rotor
    balancing, 155, *156*, 157
    dynamics, 25–26, *28*, *29*
    faults, **126**, 144; *see also* rotor faults detection
        misalignment, *127*
    rig speed profile, *215*
    with single plane unbalance, *30*
    solid point on, 30
    unbalance estimation, *155*
    unbalance force, 29
rotor faults detection
    mass unbalance, 125–126
    misalignment, 127
    shaft bow/bend, 126–127
    shaft crack, 128
    shaft rub, 128–129
running speeds/GMFs, 149, **150**
run to failure, 4

Index

237

S

sampling frequency, 61, *61*, **62**
  at 2/1 kHz, 750 Hz, 63–64, *64*
  at 10/5 kHz, 63, *63*
  at 500/400 Hz, 65, *65*
  collected signals, *62*
  sine wave, *63*
SDOF, *see* single degree of freedom (SDOF)
    system
seismometer, 47
sensitivity method, 153
sensors and mounting approach, 45
  acceleration, 48–53
  displacement, 45–46
  tacho, *53*, 53–54, *54*, *55*
  velocity, 47–48, *48*
shaft bow/bend, 126–127
shaft centerline plot, 118, *119*
shaft crack, 128
shaft misalignment, *146*
shaft response, *138*
shaft rubs, 128–129
  conditions, *146*
  simulation, *147*
short time Fourier transformation (STFT),
    100–101
  analyses, 198–199, *201*
  experimental rig, 101
  industrial centrifugal pump, 101–103,
    102, *103*
signal processing
  correlation between signals, 103–108
  FT, 82–92
  PSD, computation, 92–98
  STFT, 100–103
  time signal, 73–82, *92*
signal waveforms, 80, *82*
sine waveform, *89*, *91*
single degree of freedom (SDOF) system, 11,
    22–23, *162*
  and behavior, *12*
  damped, *14*
  under damped, *18*
  with natural period, oscillation, *13*
  responses, **22**
  on vibrating object, *49*
single plane balancing, 153
  graphical approach, *153*, 153–155
  mathematical approach, 155–157
single stage gearbox, 130
small instrumented hammer,
    photograph, *160*
spectrum analysis, 194, 212
  frequency resolution in, *87*,
    87–88, *88*
  measured vibration, *196*

spectrum/spectra
  acceleration, *see* acceleration spectrum
  amplitude, 85, 94, 162, *173*
  coherence composite, 230
  cross power, 103–104
  displacement, *100*
  gearbox, *133*, *134*
  IAS, *217*
  measured force, *163*
  normal operation condition, 113, *114*, 115
  plots, *85*, *86*, *87*, *95*, *205*
  pulse train waveform, *212*
  3D waterfall plot, 117–118, *118*
  time domain signal and, *140*
  velocity, 99, *100*
  vibration, *130*
  waterfall plot, *101*
spring-mass system, 11–12, 23, *46*
steam TG set, 198–201; *see also* turbo-generator
    (TG) set
STFT, *see* short time Fourier transformation
    (STFT)

T

tacho sensors, *53*, 53–54, *54*, *55*
tacho signal, 53–54, 153, *154*
temperature monitoring, 1, 3
TG, *see* turbo-generator (TG) set
thermography, 3
3-D waterfall plot, 117
time domain analyses, 222
time domain signal, 73, *140*
  integration, 76–78, *77*
  into segments, *93*
  and spectrum, *140*
  using STFT analysis, *101*
time domain zero-crossing approach, 210–212, 214
time-frequency analyses, 222
time signal, 73
  CF *versus* Ku, 79–82, *81*, **82**
  filters, 73–74, *75*
  statistical parameters, 78–79, *79*, **80**
  time domain signal, integration, 76–78, *77*
  vibration amplitude, 74–76, *76*
time wave-forms, *161*
torsional vibration, 209
transient operation conditions
  Bode plot, 119, *120*
  orbit plot, 118
  shaft centerline plot, 118, *119*
  spectra, 3D waterfall plot, 117–118, *118*
transient response, 19
transition piece (TP), 188
trial-and-error approach, 36
trial run, 154
turbo-generator (TG) set, 55, *56*, 198, *201*

238 Index

U

under damped system, 16–17, *18*
Upper LTM (ULTM), 7

V

VCM, *see* vibration-based condition monitoring (VCM)
velocity sensors, 47–48, *48*
velocity spectra, 99, *100*
vertical centrifugal pump, 186–188
vibration acceleration, 193, *195*
  spectrum plots, *205*
vibration-based condition monitoring (VCM), 11, 32, *122*, 225, *226*
  data processing in, *226*
  limitations in, 225, 227
  machine, 38–39
  measurement interval, 40–41
  measurement locations and directions, 39, *39*, *40*
  procedure, *38*
  software and instrumentations, 40
  vibration severity limits, **41**, 41–42
vibration data fusion concept, 229, *229*
vibration data presentation formats
  normal operation condition, 113–117
  transient operation conditions, 117–120

vibration displacement
  signal, *141*, *154*
  velocity *versus* acceleration spectra, *123*, *124*
vibration measurements and monitoring stages
  ADC, 67–69
  aged machine, 42
  aliasing affect and anti-aliasing filter, 61, 63–67, *66*, *67*
  bath tub concept, 35
  data collection steps, 57–58
  instrument calibration and specifications, 59–60
  machine installation and commissioning, 35–37, *37*
  machine operation, 38–42
  Nyquist frequency, 66–67
  sampling frequency, 61, *61*, *62*, **62**
  setup, 55–57
vibration monitoring, 3
vibration severity limits, **41**, 41–42
vibration spectrum, *130*
vibration transducer, *46*
vibration velocity signal, *140*

W

wind turbine, 188, *189*, *190*, **191**

Z

zebra reflective strip, 209, *210*